Lecture Notes in Mathematics

Edited by A. Dold and B. Eckmann

T0253961

555

Makoto Ishida

The Genus Fields of Algebraic Number Fields

Springer-Verlag
Berlin · Heidelberg · New York 1976

Author
Makoto Ishida
Department of Mathematics
Tokyo Metropolitan University
Fukazawa, Setagaya
Tokyo 158/Japan

Library of Congress Cataloging in Publication Data

Ishida, Makoto, 1932-
The genus fields of algebraic number fields.

(Lecture notes in mathematics ; 555)
Bibliography: p.
Includes index.
1. Fields, Algebraic. 2. Class field theory.
I. Title. II. Series: Lecture notes in mathematics (Berlin) ; 555.
QA3.L28 no. 555 [QA247] 510'.8s [512'.74] 76-49479

AMS Subject Classifications (1970): 12A25, 12A30, 12A35, 12A40, 12A50, 12A65

ISBN 3-540-08000-7 Springer-Verlag Berlin · Heidelberg · New York
ISBN 0-387-08000-7 Springer-Verlag New York · Heidelberg · Berlin

This work is subject to copyright. All rights are reserved, whether the whole or part of the material is concerned, specifically those of translation, re-printing, re-use of illustrations, broadcasting, reproduction by photocopying machine or similar means, and storage in data banks.

Under § 54 of the German Copyright Law where copies are made for other than private use, a fee is payable to the publisher, the amount of the fee to be determined by agreement with the publisher.

© by Springer-Verlag Berlin · Heidelberg 1976

Printed in Germany

Printing and binding: Beltz Offsetdruck, Hemsbach/Bergstr.

Preface

These notes are taken from the lectures on algebraic number theory which I have given at several universities in Japan (Tokyo University of Education, Nagoya University, Kyushu University, Hokkaido University and Ochanomizu University) and include also the results obtained thereafter.

* * *

The genus theory (Theorie der Geschlechter) of quadratic number fields has its origin in 'Disquisitiones Arithmeticae' of Gauss (cf. Cohn [3]). In 1951, Hasse [12] gave a class field theoretical interpretation of the theory. Here we quote from his introduction : 'Ist man einmal im Besitz der Hauptsätze der Klassenkörpertheorie, so lässt sich die Geschlechtertheorie ganz einfach in durchsichtiger, rein begrifflicher Gestalt herausarbeiten.'

Later on, the cyclic case was treated in Iyanaga and Tamagawa [22] and the abelian case in Leopoldt [23]. We also quote from the introduction in [23] : 'Die Aufgabe der Theorie ist es, aus den arithmetischen Eigenschaften von K der Geschlechterkörper K^* explizit zu bestimmen, dessen Relativgrad $g^+ = (K^* : K)$ ___ die Anzahl der Geschlechter ___ im Mittelpunkt der älteren Theorie stand.'

Then in 1959, Fröhlich [6], [7] generalized the notion of

the genus fields of algebraic number fields to not necessarily abelian case and obtained several interesting results on algebraic number fields of certain type. After him, some contributions have been made to the theory by Furuta [8], Madan [24], Frey and Geyer [5] and Ishida [15], [16], [18], [19], [20].

* * *

Chapter 1 is of preliminary nature and there is given the definition of the genus fields. We consider algebraic number fields of Eisenstein type in Chapter 2 and show several results on class numbers. In Chapter 3, we construct an unramified abelian extension over a given algebraic number field K as a composite of K with an absolute abelian number field (in 'wide' sense as well as in 'narrow' sense). In Chapter 4, we define the genus field K^* and the genus number g_K of an algebraic number field K according to Fröhlich. An explicit method of constructing the genus field is given there. The classical theorem on quadratic number fields and the theorem of Leopoldt on genus numbers of absolute abelian number fields are also reproduced there. In Chapter 5, we treat the case where the degree of K is an odd prime number. In particular, in Chapter 6, the genus field K^* of a cubic number field K is explicitly determined by the minimal polynomial of a primitive element of K. In Chapter 7, we determine the genus field K^* of a pure number field of odd degree.

*　　　*　　　*

Some parts of this note were completed when I was suffering from a kidney disease. During that period, I was greatly helped and encouraged by the kindness of many people, to whom I wish to express my hearty gratitude. In particular, I am very grateful to my colleagues at the Department of Mathematics, Tokyo Metropolitan University, to Professor N.Akiyama of the Institute of Medical Science, University of Tokyo, to Doctors and Nurses of the Hemo-Dialysis Center and of the 5th Ward, Tokyo Central Hospital of Social Health Insurance, and to my wife Hiroko and my two daughters Kumiko and Tomoko.

October, 1975

M. I.

* * *

Some parts of this note were completed when I was suffering from a kidney disease. During that period, I was greatly helped and encouraged by the kindness of many people, to whom I wish to express my hearty gratitude. In particular, I am very grateful to my colleagues at the Department of Mathematics, Tokyo Metropolitan University, to Professor N.Akiyama of the Institute of Medical Science, University of Tokyo, to Doctors and Nurses of the Hemo-Dialysis Center and of the 5th Ward, Tokyo Central Hospital of Social Health Insurance, and to my wife Hiroko and my two daughters Kumiko and Tomoko.

October, 1975

M. I.

Table of Contents

Table of Contents

Chapter 1. Preliminaries

Let K be an algebraic number field of finite degree. Then it is an important problem in algebraic number theory to investigate the ideal class group or the class number of K (in 'wide' sense as well as in 'narrow' sense). As is well known in class field theory, this problem is closely connected with the study of unramified abelian extensions of K. Clearly, among such extensions, those which are obtained as a composite of K with absolute abelian number fields seem to be approachable. As for such abelian extensions of K, Fröhlich has given the following definition :

Definition (Fröhlich [6]). Let K be an algebraic number field. Then the genus field K^* of K is the maximal abelian extension of K, which is composed of an absolute abelian number field k^* with K and is unramified at all the finite prime ideals of K. The extension degree $g_K = [K^* : K]$ is called the genus number of K. The Galois group $\mathfrak{G}$ of K^* over K is also called the genus group of K.

In terms of class field theory, the genus field K^* of K is the abelian extension corresponding to an ideal group $G(K)$, in K, which contains the group of all totally positive principal ideals. Denoting by $A(K)$ the group of all ideals of K, we know that the factor group $A(K)/G(K)$ is isomorphic to the Galois group of K^* over K i.e. the genus group of K and so the index $(A(K) : G(K))$ is equal to the genus number g_K of K. The ideal group $G(K)$ is called the <u>principal genus</u> of K.

Let k^* be the maximal absolute abelian subfield of K^*. Then we have also $K^* = k^*K$. In this note, our main interest is concerned with this absolute abelian number field k^* rather than K^*.

The situation of Fröhlich's definition is clarified when we see the following diagram :

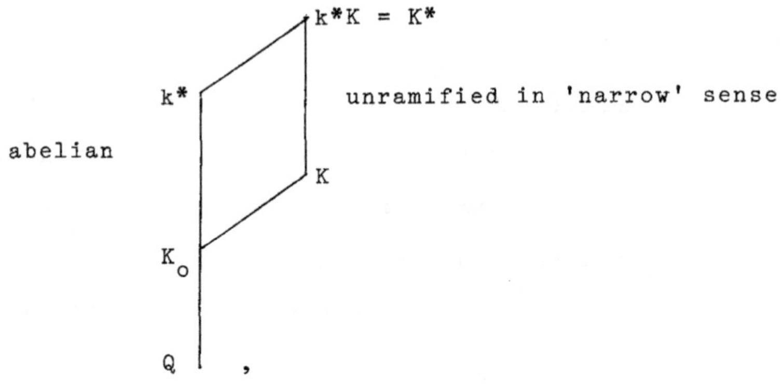

where K_o is the maximal abelian subfield of K and so we have

$$g_K = [k^* : Q] / [K_o : Q].$$

Moreover, the genus group $\mathcal{G}$ of K is isomorphic to the Galois group of k^* over K_o.

In case K is an absolute abelian number field, this definition coincides with that of Leopoldt [23] ; Leopoldt has defined, for an absolute abelian number field K, the <u>genus field</u> K^* of K as the maximal absolute abelian number field containing K, which is unramified at all the finite prime ideals of K. The above diagram is reduced to the following form in this case :

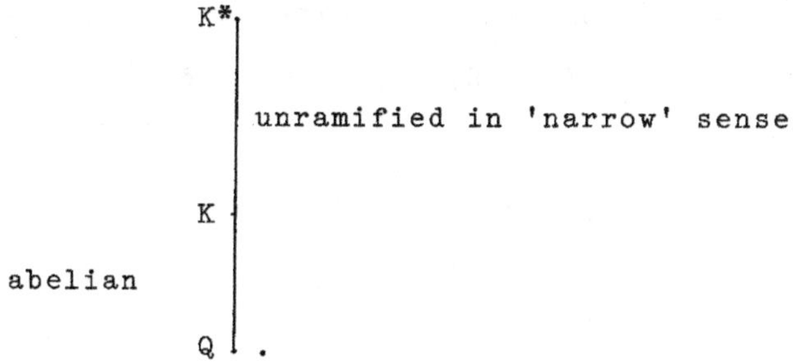

Clearly, in this case, we have $K^* = k^*$.

In fact, this is the definition of Hasse [12] for the quadratic number fields and there he proved the classical theorem of Gauss in terms of class field theory. In order to clarify our purpose, we state here some results of Hasse (cf. Cohn [3]).

Let $K = Q(\sqrt{m})$ be a quadratic number field, where m is a squarefree integer. The discriminant D_K of K is given by

$$(1.1)\qquad D_K = \begin{cases} m, & \text{if } m \equiv 1 \pmod 4, \\ 4m, & \text{if } m \equiv 2 \text{ or } 3 \pmod 4. \end{cases}$$

Moreover, for a prime divisor p of m , we define the prime (quadratic) discriminant p^* as follows :

$$(1.2)\qquad p^* = \begin{cases} (-1)^{(p-1)/2}\, p \ \text{ i.e. } p \text{ or } -p \text{ according} \\ \qquad \text{to } p \equiv 1 \text{ or } 3 \pmod 4, \text{ if } p \text{ is odd}, \\ -4,\ 8 \text{ or } -8 \text{ according to } m \equiv 3 \pmod 4, \\ \qquad 2 \pmod 8 \text{ or } -2 \pmod 8, \text{ if } p = 2, \end{cases}$$

which is the discriminant of the quadratic number field $Q(\sqrt{p^*})$. If $p_1, p_2, \ldots, p_t$ are all the distinct prime divisors of D_K, we have

$$(1.3)\qquad D_K = p_1^* \, p_2^* \cdots p_t^*.$$

Then the theorem of Hasse shows that the genus field $K^* = k^*$ of K is given by

$$(1.4)\qquad \begin{aligned} k^* &= Q(\sqrt{p_1^*}, \sqrt{p_2^*}, \ldots, \sqrt{p_t^*}), \\ K^* &= K(\sqrt{p_1^*}, \sqrt{p_2^*}, \ldots, \sqrt{p_t^*}). \end{aligned}$$

Here, as $\sqrt{p_1^*}\sqrt{p_2^*}\ldots\sqrt{p_t^*} = \sqrt{D_K}$ is in K, we see $K^* = k^*$. Thus the genus field K^* of K is determined explicitly by the arithmetic properties of K. Moreover the genus number g_K of K is given by

(1.5) $$g_K = 2^{t-1}.$$

Since the Galois group $\mathfrak{G}$ of K^* over K (the genus group of K) is an abelian group of order 2^{t-1} and of type $(2,2,\ldots,2)$ and the ideal class group C_K^+ of K in 'narrow' sense has a factor group isomorphic to $\mathfrak{G}$, we see that the 2-rank $d^+ = d^{(2)} C_K^+$ of C_K^+ satisfies

(1.6) $$d^+ \geqq t - 1.$$

A further consideration shows that, in (1.6), the equality always holds i.e. we have

(1.6)' $$d^+ = t - 1.$$

As is easily expected by the definition (or by the above diagram), the so-called 'Verschiebungssatz' in class field theory plays an essential role in our investigations.

The 'Verschiebungssatz' (Hasse [10], Takagi [26]). Let K be an algebraic number field of finite degree and let k be an absolute abelian number field. Suppose that k corresponds to the ideal group $H_{\tilde{m}}$, in Q, with defining modulus $\tilde{m}$. Then the abelian extension kK of K corresponds to the ideal group $H_{\tilde{m}}(K)$, in K, with defining modulus $\tilde{m}$ such that

$$H_{\tilde{m}}(K) = \left\{ \mathfrak{a} : \text{ideals of } K \;\middle|\; (\mathfrak{a}, \tilde{m}) = 1 \text{ and } N_K \mathfrak{a} \in H_{\tilde{m}} \right\}.$$

Since the maximal abelian extension $\bar{K}$ of K, which is unramified at all the finite prime ideals, corresponds to the group of all totally positive principal ideals in K, we have the following

Corollary. In the notations of Theorem, the abelian extension kK is unramified at all the finite prime ideals of K i.e. $kK \subset \bar{K}$ if and only if $H_{\tilde{m}}(K)$ contains all the totally positive principal ideals (γ), prime to $\tilde{m}$, in K i.e. $N_K(\gamma) \in H_{\tilde{m}}$.

Besides this theorem, the following two well-known classical theorems are also essential.

<u>Theorem of Minkowski</u>. If K is an algebraic number field other than Q, then we have $|D_K| > 1$ i.e. there is at least one rational prime number ramifying in K.

<u>Theorem of Kronecker and Weber</u>. Let k be an absolute abelian number field. Then k is contained in some cyclotomic number field $Q(\zeta_m)$ $(m \in Z)$. Moreover, we can take m so that, for a rational prime number p, p is ramified in k if and only if $p \mid m$.

We use these Theorems and Corollary freely without reference.

Moreover the following remarks are needed. Let

$$(1.7) \qquad (p) = \mathfrak{p}_1^{e_1} \mathfrak{p}_2^{e_2} \cdots \mathfrak{p}_m^{e_m}$$

be the prime ideal decomposition of p in K, where $\mathfrak{p}_i$ are prime ideals of K and $N_K \mathfrak{p}_i = p^{f_i}$. Then we have

$$(1.8) \qquad \sum_{i=1}^{m} e_i f_i = [K : Q].$$

In particular, if we denote by $e(p)$ the greatest common divisor of $e_1, e_2, \dots, e_m$, then $e(p)$ is a divisor of $[K : Q]$;

(1.9) $$e(p) \mid [K : Q].$$

On the other hand, suppose that k is an absolute abelian number field such that kK is unramified at all the finite prime ideals of K. Then, for any rational prime number p, the multiplicative property of the ramification indices implies that the ramification index e^* of p in k divides $e_1, e_2, \dots, e_m$ i.e. we have

(1.10) $$e^* \mid e_1, e_2, \dots, e_m \quad \text{and so} \quad e^* \mid e(p).$$

Chapter 2. Algebraic number fields of Eisenstein type

Let K be an algebraic number field of degree n over Q. We say that K is of Eisenstein type with respect to a rational prime number p, if K is obtained by adjunction, to Q, a root α of an Eisenstein polynomial with respect to p ; that is, there is a primitive element α of K, which is an integer of K and whose minimal polynomial is

$$(2.1) \qquad f(X) = X^n + a_1 X^{n-1} + \dots + a_n \in Z[X]$$

with $p \mid a_i$ and $p \| a_n$. Then we see easily that

$$(2.2) \qquad (p) = \mathfrak{p}^n \quad \text{in } K \quad \text{and} \quad \mathfrak{p} \| \alpha ,$$

where $\mathfrak{p}$ is a prime ideal of K (cf. Borevich and Shafarevich [1]).

Conversely suppose that, for an algebraic number field K of degree n, we have $(p) = \mathfrak{p}^n$ in K, where p is a rational prime number and $\mathfrak{p}$ is a prime ideal of K. Then there exists an integer α in K such that $\mathfrak{p} \| \alpha$ and we have $K = Q(\alpha)$. In fact, if $K_0 = Q(\alpha) \subsetneqq K$, there is a prime ideal $\mathfrak{p}_0$, of K_0, dividing p and α. Then clearly $\mathfrak{p}^{[K : K_0]}$ divides α , which is a contradiction. Let $f(X) = X^n + a_1 X^{n-1} + \dots + a_n \in$

Z[X] be the minimal polynomial of α over Q. As p is totally ramified in K, we have, by a well-known theorem (cf. Takagi [26]),

$$f(X) \equiv (X - t)^n \pmod{p}$$

with $t \in Z$; and so $p \mid a_1, p \mid a_2, \ldots, p \mid a_n$. Comparing the maximal exponent of $\mathfrak{p}$ in each term of $f(\alpha) = 0$, we have $p \mid a_n$. In fact, these maximal exponents are $n, \geqq n + (n-1), \geqq n + (n-2), \ldots, \geqq n + 1, ne$ respectively, where $p^e \parallel a_n$ $(e > 0)$. Then, as at least two of them take the smallest value, we have $n = ne$ and $e = 1$. Hence $f(X)$ is an Eisenstein polynomial with respect to p and so K is of Eisenstein type with respect to p.

Thus we have the following

<u>Proposition 1</u>. For an algebraic number field K, the following two assertions are equivalent.

(1) K is of Eisenstein type with respect to p,

(2) p is totally ramified in K.

<u>Corollary</u>. If K is of Eisenstein type with respect to p, then so is any subfield of K.

Now we consider an algebraic number field K of Eisenstein

type with respect to p and of degree n over Q. Let $K = Q(\alpha)$, where (2.1) is the minimal polynomial of α over Q. (A typical example of algebraic number fields of Eisenstein type is the p-th cyclotomic number field $Q(\zeta_p)$ for a prime p. In this case, the following Lemmas are well-known.)

<u>Lemma 1</u>. Let O_K be the ring of integers in K. Then we have

$$(2.3) \qquad p \nmid (O_K : Z[\alpha]).$$

<u>Proof</u>. If $p \mid (O_K : Z[\alpha])$, there is an element ω of O_K such that

$$\omega \notin Z[\alpha] \quad \text{and} \quad p\omega = \sum_{i=0}^{n-1} x_i \alpha^i \in Z[\alpha]$$

with $x_i \in Z$. Then we have, by (2.2),

$$\begin{aligned} p\omega = \sum_{i=0}^{n-1} x_i \alpha^i \in \mathfrak{p}^n &\Longrightarrow x_0 \in \mathfrak{p} \Longrightarrow x_0 \in (p) = \mathfrak{p}^n \\ \Longrightarrow \sum_{i=1}^{n-1} x_i \alpha^i \in \mathfrak{p}^n &\Longrightarrow x_1 \alpha \in \mathfrak{p}^2 \Longrightarrow x_1 \in \mathfrak{p} \\ \Longrightarrow x_1 \in (p) = \mathfrak{p}^n &\Longrightarrow \cdots \Longrightarrow x_{n-1} \in (p) \end{aligned}$$

and so we have $\omega \in Z[\alpha]$, which contradicts the assumption. #

Lemma 2. For any integer γ in K, we have

$$N_K \gamma \equiv x^n \pmod{p} \tag{2.4}$$

with some $x \in Z$.

Proof. Let $c = (O_K : Z[\alpha])$; then we have, by Lemma 1, $p \nmid c$ and $c\gamma = \sum_{i=0}^{n-1} x_i \alpha^i \in Z[\alpha]$ $(x_i \in Z)$. Since $-a_1, a_2, \ldots, (-1)^n a_n$ are fundamental symmetric polynomials of all the conjugates of α over Q and so we have $N_K c\gamma = c^n N_K \gamma = x_o{}^n + \{$ Z-linear combination of monomials of $a_1, a_2, \ldots, a_n \}$, we have $c^n N_K \gamma \equiv x_o{}^n \pmod{p}$ with $p \nmid c$. #

The result stated above holds in local case as follows : Let p be a rational prime number and $F = Q_p$ the p-adic completion of Q. Then p is a prime element of F. Consider a finite extension E of F with the ramification index e and the residue class degree f ; so we have $[E : F] = ef$. Let Π be a prime element of E. Then it is known that there is an intermediate field T of E and F with the following properties (cf. Hasse [11]) :

i) T is an unramified extension, of degree f, of F and so p is also a prime element of T.

ii) E is a totally ramified extension, of degree e, of T.

iii) We have $E = T(\Pi)$ and the minimal polynomial of Π over T is an Eisenstein polynomial with respect to p.

iv) $1, \Pi, \ldots, \Pi^{e-1}$ constitute the integral basis of E over T.

Hence any integer Γ of E can be written as

$$\Gamma = \beta_o + \beta_1\Pi + \ldots + \beta_{e-1}\Pi^{e-1}$$

with integers β_i in T ; and so, in a similar way as in the proof of Lemma 2, we have

$$N_{E/T}\Gamma = \beta_o{}^e + p\,\delta$$

with an integer δ in T. As p is in F, it follows

$$N_{E/F}\Gamma = N_{T/F}\,N_{E/T}\Gamma = (N_{T/F}\beta_o)^e + pd$$

with an integer d in F. So we have the following

<u>Lemma 3</u>. For any integer Γ in E, we have

$$N_{E/F}\Gamma = c^e + pd$$

with some integers c, d in F.

Now using the results on local case (Lemma 3), we can generalize Lemma 2 as follows : Let K be an algebraic number field of finite degree n. Suppose that

$$(2.5) \qquad (p) = \mathfrak{p}_1^{e_1}\,\mathfrak{p}_2^{e_2} \cdots \mathfrak{p}_m^{e_m} \quad \text{in } K,$$

$$N_K \mathfrak{p}_i = p^{f_i} \quad \text{and} \quad \sum_{i=1}^{m} e_i f_i = n,$$

where $\mathfrak{p}_1, \mathfrak{p}_2, \ldots, \mathfrak{p}_m$ are distinct prime ideals of K ($e_i > 0$). Let $E_i = K_{\mathfrak{p}_i}$ be the $\mathfrak{p}_i$-adic completion of K. Then E_i is an extension of $F = Q_p$ with the ramification index e_i and the residue class degree f_i. On the other hand, it is known that we have, for any number β of K,

$$N_K \beta = \prod_{i=1}^{m} N_{E_i/F}\, \beta$$

(cf. Hasse [11]). Hence, by Lemma 3, we have, for any integer γ in K,

$$N_K \gamma = \prod_{i=1}^{m} N_{E_i/F}\, \gamma = \prod_{i=1}^{m} (c_i^{e_i} + p d_i),$$

where c_i, d_i are integers in F. As any integer in F is congruent to an integer in Z modulo p, we have the following

Lemma 4 (Ishida [18]). Let $e = (e_1, e_2, \ldots, e_m)$. Then, for any integer γ in K, we have

$$(2.6) \qquad N_K \gamma \equiv c^e \pmod{p}$$

with some $c \in Z$.

Here we note that the equality $\sum_{i=1}^{m} e_i f_i = n$ implies that e is a divisor of $n = [K : Q]$.

Appendices to Chapter 2

Using Proposition 1 and Lemma 2, which are of elementary nature, we can prove some properties of class numbers of algebraic number fields in elementary way.

A) (Ishida [14], [21]) Let q be an odd prime number and let K be an algebraic number field of degree q. We denote by C_K the ideal class group of K and by $d = d^{(q)} C_K$ the q-rank of C_K i.e.

$$q^d = (C_K : C_K{}^q).$$

Let $p_1, p_2, \ldots, p_t$ be all the rational prime numbers which are totally ramified in K : $(p_i) = \mathfrak{p}_i^{\,q}$. Then we prove the inequality

$$(2.7) \qquad d = d^{(q)}\, C_K \geqq t - q \quad \text{and so} \quad q^{t-q} \,\big|\, h_K,$$

where h_K is the class number of K. In fact, more precisely, we prove

$$(2.8) \qquad d = d^{(q)}\, C_K \geqq t - r - 1,$$

where $r = r_1 + r_2 - 1$ and r_1, $2r_2$ denote the numbers of real and complex conjugates of K over Q respectively ($r_1 + 2r_2 = q$).

Now suppose that, on the contrary to (2.8), we have the inequality

$$d < t - r - 1 \quad \text{i.e.} \quad t - 1 \geqq d + r + 1.$$

Then, as $t \geqq 2$, we may assume $p_t \neq q$ (by changing the index if necessary). Let $\mathcal{L}_i$ be the element of C_K (the ideal class) containing $\mathfrak{p}_i$; so, as $\mathfrak{p}_i^{\,q} = (p_i)$, we have $\mathcal{L}_i^{\,q} = 1$. Hence, by the definition of $d = d^{(q)}\, C_K$ ($d + 1 \leqq t - 1$), there are $x_1, x_2, \ldots, x_{d+1} \in Z$ not all divisible by q such that

$$(2.9)_1 \qquad \mathfrak{L}_1^{x_1}\mathfrak{L}_2^{x_2} \cdots \mathfrak{L}_{d+1}^{x_{d+1}} = 1,$$

$$\text{i.e.} \quad \mathfrak{p}_1^{x_1}\mathfrak{p}_2^{x_2} \cdots \mathfrak{p}_{d+1}^{x_{d+1}} = (\mu_1)$$

with $\mu_1 \in O_K$. Here we may also assume $q \nmid x_1 = x_1^{(1)}$ (by changing the index if necessary). Applying the same reasoning to $\mathfrak{p}_2, \mathfrak{p}_3, \ldots, \mathfrak{p}_{d+2}$ $(d + 2 \leqq t - 1)$, we may have

$$(2.9)_2 \qquad \mathfrak{p}_2^{x_2^{(2)}}\mathfrak{p}_3^{x_3^{(2)}} \cdots \mathfrak{p}_{d+2}^{x_{d+2}^{(2)}} = (\mu_2),$$
$$x_j^{(2)} \in Z, \quad q \nmid x_2^{(2)}$$

with $\mu_2 \in O_K$ (by changing the index if necessary). We continue these processes and we have, for $i = 1, 2, \ldots, r+1$ $(d + r + 1 \leqq t - 1)$,

$$(2.9)_i \qquad \mathfrak{p}_i^{x_i^{(i)}}\mathfrak{p}_{i+1}^{x_{i+1}^{(i)}} \cdots \mathfrak{p}_{d+i}^{x_{d+i}^{(i)}} = (\mu_i),$$
$$x_j^{(i)} \in Z, \quad q \nmid x_i^{(i)}$$

with $\mu_i \in O_K$ (by changing the index if necessary). Taking the q-th power of each equality $(2.9)_i$, we have the following relations between principal ideals :

$$(p_i)^{x_i^{(i)}}(p_{i+1})^{x_{i+1}^{(i)}} \cdots (p_{d+i})^{x_{d+i}^{(i)}} = (\mu_i^q).$$

Let $\mathcal{E}_1, \mathcal{E}_2, \ldots, \mathcal{E}_r$ $(r = r_1 + r_2 - 1)$ be the system of fundamental units of K. As K has no root of unity other than ± 1, it follows

$$(2.10)_i \qquad p_i^{x_i^{(i)}} p_{i+1}^{x_{i+1}^{(i)}} \cdots p_{d+i}^{x_{d+i}^{(i)}} = \pm \mathcal{E}_1^{y_1^{(i)}} \mathcal{E}_2^{y_2^{(i)}} \cdots \mathcal{E}_r^{y_r^{(i)}} \mu_i^{q}$$

with $y_j^{(i)} \in Z$ for $i = 1, 2, \ldots, r+1$. Then linear homogeneous equations

$$y_j^{(1)} z_1 + y_j^{(2)} z_2 + \cdots + y_j^{(r+1)} z_{r+1} = 0$$

for $j = 1, 2, \ldots, r$ have clearly a non-trivial solution $z_1 = a_1, z_2 = a_2, \ldots, z_{r+1} = a_{r+1}$ in Z such that at least one of a_i is not divisible by q. Also taking the a_i-th power of each equality $(2.10)_i$, we have

$$(2.11) \qquad \prod_{i=1}^{r+1} (p_i^{x_i^{(i)}} p_{i+1}^{x_{i+1}^{(i)}} \cdots p_{d+i}^{x_{d+i}^{(i)}})^{a_i} = (\pm \prod_{i=1}^{r+1} \mu_i^{a_i})^q.$$

On the other hand, supposing $q \mid a_1, \ldots, q \mid a_{i-1}, q \nmid a_i$, we see that, in the left side of (2.11), the exponent of p_i is not divisible by q. That is, $A = \prod_{i=1}^{r+1} (\prod_{j=0}^{d} p_{i+j}^{x_{i+j}^{(i)}})^{a_i}$ is not equal to the q-th power of any number in Q. Consequently, by (2.11), $\alpha = \pm \prod_{i=1}^{r+1} \mu_i^{a_i}$ is not in Q and we have

$$Q \subsetneqq Q(\alpha) \subset K \quad \text{and so} \quad Q(\alpha) = K.$$

Since the minimal polynomial $X^q - A$ of α has the discriminant $D_\alpha = \pm q^q A^{q-1}$, all the rational prime numbers ramifying in K are contained in the set $\{ q, p_1, p_2, \ldots, p_{d+r+1} \}$. However this contradicts the assumption that p_t ($\neq q$ and $t > d + r + 1$) is ramified in K. Hence we have (2.8) and, as $r + 1 = r_1 + r_2 \leqq q$, also (2.7).

In the above reasoning, if K is assumed to be not equal to a pure number field $Q(\sqrt[q]{A})$ $(A \in Q)$, then, in the proof of (2.8), we need not use the existence of a ramifying prime p_t. Hence, in this case, we have a more strict inequality

$$(2.12) \qquad d = d^{(q)}_{C_K} \geqq t - r \geqq t - q + 1.$$

By (2.7), we can, for a given odd prime number q and a given natural number e, construct infinitely many algebraic number fields K of degree q such that

(2.13) $\quad d^{(q)} C_K \geqq e \quad$ and so $\quad q^e \mid h_K$.

In fact, let $t = q + e$ and take t rational prime numbers $p_1, p_2, \ldots, p_t$. We construct an algebraic number field K of degree q and of Eisenstein type with respect to each p_i $(i = 1, 2, \ldots, t)$. (For example, adjoin a root of $X^q + p_1p_2\cdots p_t$ or $X^q + p_1p_2\cdots p_tX + p_1p_2\cdots p_t$ to Q.) Then as, by Proposition 1, each p_i is totally ramified in K, (2.7) implies $d^{(q)} C_K \geqq t - q = e$ and so $q^e \mid h_K$. Now suppose that there are given algebraic number fields $K_1, K_2, \ldots, K_s$ such that $[K_i : Q] = q$ and $d^{(q)} C_{K_i} \geqq e$. Take a rational prime number p not dividing the discriminant D_{K_i} of each K_i $(i= 1, 2, \ldots, s)$ and construct, by the method stated above, an algebraic number field K for $p_1, p_2, \ldots, p_t = p$ such that $[K : Q] = q$ and $d^{(q)} C_K \geqq e$. As $p = p_t$ is totally ramified in K, we have $K \neq K_1, K_2, \ldots, K_s$ and so, in this way, we can construct infinitely many algebraic number fields K, of degree q, satisfying (2.13).

These results (2.7) and (2.8) containing (2.13) is a slight generalization of a well-known theorem of quadratic case (cf. Chapter 1) to the case of odd prime degree. (In quadratic case, we treat the ideal group C_K^+ in 'narrow' sense in place of C_K.)

A similar inequality for general algebraic number fields is obtained in Roquette and Zassenhaus [25].

Example 1. For $q = 5$, let $K = Q(\sqrt[5]{2.3.5.7.11.13})$. Then we have, by (2.7), $d^{(5)}C_K \geq 6 - 5 = 1$ and so $5 \mid h_K$. More precisely, as $r_1 = 1$, $r_2 = 2$ and $r = 2$, we have, by (2.8), $d^{(5)}C_K \geq 6 - 2 - 1 = 3$ and so $5^3 \mid h_K$. #

B) (Ishida [15], [21]) Let n be a natural number (> 1) and p a rational prime number such that $p \equiv 1 \pmod{2n}$. Let K be an algebraic number field of degree n and of Eisenstein type with respect to p.

For an integral ideal $\mathfrak{a}$, prime to p, m denotes the order of the element of C_K (the ideal class) containing $\mathfrak{a}$. Here we note that m is a divisor of the class number h_K of K. Then, as $\mathfrak{a}^m = (\gamma)$ with $\gamma \in O_K$, we have, by Lemma 2,

$$(2.14) \qquad a^m = N_K \mathfrak{a}^m = \left| N_K \gamma \right| \equiv \pm x^n \pmod{p},$$

where $a = N_K \mathfrak{a}$ and $x \in Z$. Let f be the order of $a = N_K \mathfrak{a}$ modulo p (i.e. f is the smallest natural number such that $a^f \equiv 1 \pmod{p}$). Taking the $(p-1)/n$-th power of (2.14), we have $a^{m(p-1)/n} \equiv 1 \pmod{p}$ and so f must divide $m(p-1)/n$. So n is a divisor of $m(p-1)/f$. Hence, if $(n, (p-1)/f) = 1$, then we have $n \mid m$. Therefore, in this case, C_K contains an element of order n and h_K is divisible by n.

For example, suppose that K is of degree n (>1) and of Eisenstein type with respect to two rational prime numbers p and q, where $p \equiv 1 \pmod{2n}$. Then, as q is totally ramified in K, there is a prime ideal $\mathfrak{q}$ of K with $N_K \mathfrak{q} = q$. Denoting by f the order of q modulo p, we see that $(n, (p-1)/f) = 1$ implies $n \mid h_K$. In particular, if q is a primitive root modulo p, then we have $f = p-1$ and so n is a divisor of h_K.

Example 2. For $n = 5$, let $p = 11 \equiv 1 \pmod{10}$, $q = 2$ and $K_1 = Q(\sqrt[5]{2.11})$. Then, as the order of 2 modulo 11 is 10 $(2^{10} \equiv 1 \pmod{11})$ and $5 \nmid 10/10$, we have $5 \mid h_{K_1}$. Also let $p = 11 \equiv 1 \pmod{10}$, $q = 3$ and $K_2 = Q(\sqrt[5]{3.11})$. Then, as the order of 3 modulo 11 is 5 $(3^5 \equiv 1 \pmod{11})$ and $5 \nmid 10/5$, we have $5 \mid h_{K_2}$. #

Chapter 3. Some unramified abelian extensions

Let K be an algebraic number field of finite degree and let p be a rational prime number. Suppose that k is an absolute abelian number field with the conductor (Führer) p such that kK is an unramified abelian extension of K. Denote $d = [k : Q]$. As is well known, such an absolute abelian number field k is contained in the maximal real subfield $Q(\zeta_p)_0$ of the p-th cyclotomic number field $Q(\zeta_p)$, which is a cyclic extension over Q and in which p is totally ramified.

Now we use the terminologies of class field theory (cf. Hasse [10], Takagi [26]). Let A_p be the group of all (principal) ideals, prime to p, and let S_p be the 'Strahl mod p' in Q i.e. the subgroup of A_p consisting of all ideals (a) with $a \equiv 1 \pmod{^\times p}$ (multiplicative congruence). Then k corresponds to an ideal group H_p with defining modulus p such that

$$A_p \supset H_p \supset S_p. \tag{3.1}$$

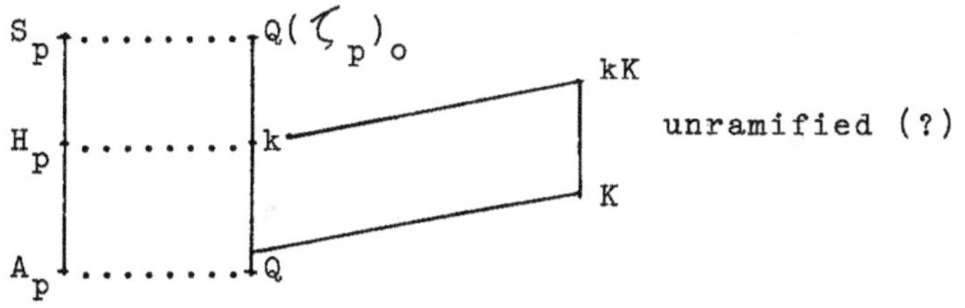

It is known that A_p/S_p is a cyclic group of order

$$(3.2) \qquad g = \begin{cases} 1 & \text{for } p = 2 , \\ (p-1)/2 & \text{for odd } p. \end{cases}$$

So H_p/S_p is the unique subgroup of A_p/S_p of index $d = [k : Q]$ and we have

$$(3.3) \qquad d \mid g.$$

On the other hand, let

$$(p) = \mathfrak{p}_1^{e_1} \mathfrak{p}_2^{e_2} \cdots \mathfrak{p}_m^{e_m}$$

in K, where $\mathfrak{p}_1, \mathfrak{p}_2, \ldots , \mathfrak{p}_m$ are distinct prime ideals of K $(e_i > 0)$. As p is totally ramified in k and each $\mathfrak{p}_i$ is unramified in kK $(i = 1, 2, \ldots , m)$, we have

$$(3.4) \qquad d \mid e_1, e_2, \ldots , e_m \quad \text{i.e.} \quad d \mid e = (e_1, e_2, \ldots , e_m).$$

Hence, by (3.3) and (3.4), we see that $d = [k : Q]$ must be a common divisor of g and e.

Conversely we can prove the following

Theorem 1 (Frey and Geyer [5], Ishida [18]). Let K be an algebraic number field of finite degree. For an odd prime number p, suppose that we have

$$(3.5)\qquad (p) = \mathfrak{p}_1^{e_1}\, \mathfrak{p}_2^{e_2} \cdots \mathfrak{p}_m^{e_m}$$

in K. Put $e = (e_1, e_2, \ldots, e_m)$ and

$$d = (e, (p-1)/2)$$

and consider the unique subfield k, of degree d, of $Q(\zeta_p)_0$. Then k gives an unramified abelian extension kK of K.

Proof. First we note that k is the absolute abelian number field corresponding to the ideal group

$$H_p = \left\{ (a) \in A_p \;\middle|\; (a)^s \in S_p \right\}, \quad s = (p-1)/2d$$

with defining modulus p in Q (which is the unique subgroup of A_p of index d and containing S_p). For a number β, prime to p, in K, Lemma 4 implies that we have

$$N_K \beta \equiv b^e \pmod{^\times p} \quad \text{i.e.} \quad b^{-e} N_K \beta \equiv 1 \pmod{^\times p}$$

with a rational number b prime to p. Then we have

$$b^{-es} N_K \beta^s \equiv 1 \pmod{^{\times} p} ;$$

so the principal ideal $(b^{-es})(N_K \beta^s)$ is in S_p. On the other hand, we have trivially

$$(b^{-es}) = (b^{-e/d})^{(p-1)/2} \in S_p .$$

Let $A_p(K)$ be the group of all ideals, prime to p, and let $P_p(K)$ be the subgroup of $A_p(K)$ consisting of all principal ideals in K respectively. Then, for any element (β) in $P_p(K)$, we see that $N_K(\beta)^s = (N_K \beta)^s$ is in S_p i.e. $N_K(\beta)$ is in H_p. Now define the ideal group $H_p(K)$ with defining modulus p in K by

$$H_p(K) = \left\{ \mathfrak{a} \in A_p(K) \,\middle|\, N_K \mathfrak{a} \in H_p \right\}.$$

Then, by the 'Verschiebungssatz', it follows that the abelian extension kK of K corresponds to $H_p(K)$. On the other hand, the fact just shown above implies that $H_p(K)$ contains $P_p(K)$ and so the ideal group $H_p(K)$ has the conductor 1. Thus Theorem 1 is proved. #

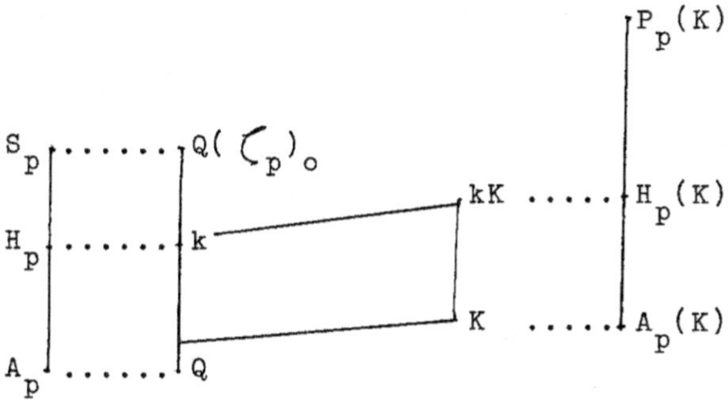

We note that p is totally ramified in k and all other rational prime numbers are unramified in k.

Remark. Theorem 1 is proved in Frey and Geyer [5] by using the following

Lemma of Abhyankar (cf. [5]). Let K_1 and K_2 be algebraic number fields with the composite $L = K_1K_2$. Let $\mathfrak{P}$ be a prime ideal of L dividing prime ideals $\mathfrak{p}_1$, $\mathfrak{p}_2$ and p of K_1, K_2 and Q respectively. Moreover let e_i be the ramification index of $\mathfrak{p}_i$ over Q. Then if $e_1 \mid e_2$ and $p \nmid e_1$ (i.e. $\mathfrak{p}_1$ is tamely ramified over Q), $\mathfrak{P}$ is unramified over K_2.

In order to treat several odd prime numbers simultaneously, we denote by k(p) the absolute abelian number field k defined in Theorem 1 with respect to p and by d(p) the degree

$[k(p) : Q]$. For distinct odd prime numbers $p_1, p_2, \cdots, p_t$, let k_1 be the composite $k(p_1)k(p_2)\ldots k(p_t)$. Then it is easily seen that k_1K is an unramified abelian extension of K and we have

$$[k_1 : Q] = d(p_1)d(p_2)\ldots d(p_t)$$

(k_1 is a subfield of $Q(\zeta_{p_1 p_2 \ldots p_t})_o$).

Moreover we can modify the above result by considering the infinite prime divisor p_∞ in Q to prove a similar result for unramified abelian extension of K in 'narrow' sense.

Put, for a rational prime number p, $\tilde{p} = pp_\infty$ and define the ideal groups $A_{\tilde{p}}$ and $S_{\tilde{p}}$ with defining modulus $\tilde{p}$ in Q similarly as in the beginning of this chapter $(A_{\tilde{p}} = A_p)$. Then an absolute abelian number field k^* with the conductor $\tilde{p}$ is contained in the p-th cyclotomic number field $Q(\zeta_p)$, which is a cyclic extension over Q and in which p is totally ramified. It is also known that A_p/S_p is a cyclic group of order

$$(3.6) \qquad g^* = \begin{cases} 1 & \text{for } p = 2, \\ p - 1 & \text{for odd } p. \end{cases}$$

Then we prove the following

Theorem 2 (Ishida [18]). The notations being as in Theorem 1, let $\tilde{p} = pp_\infty$. Put

$$d^* = (e,\ p - 1)$$

and consider the unique subfield k^*, of degree d^*, of $Q(\zeta_p)$. Then k^* gives an abelian extension k^*K of K, which is unramified at all the finite prime ideals of K.

Proof. First we note that k^* is the absolute abelian number field corresponding to the ideal group

$$H_{\tilde{p}} = \left\{ (a) \in A_{\tilde{p}} \;\middle|\; (a)^{s^*} \in S_{\tilde{p}} \right\}, \quad s^* = (p-1)/d^*$$

with defining modulus $\tilde{p}$ in Q (which is the unique subgroup of $A_{\tilde{p}}$ of index d^* and containing $S_{\tilde{p}}$). Similarly as in the proof of Theorem 1, for a number β, prime to p, in K, there is a rational number b prime to p, such that

$$b^{-es^*} N_K \beta^{s^*} \equiv 1 \quad (\mathrm{mod}^\times\, p).$$

Suppose that β is totally positive ; so $N_K \beta^{s^*}$ is positive. On the other hand, as $p - 1$ is even, we have

$$b^{-es^*} = (b^{-e/d^*})^{p-1} > 0.$$

So $(b^{-es^*})(N_K \beta^{s^*})$ is in $S_{\tilde{p}}$ and, as $(b^{-es^*}) = (b^{-e/d^*})^{p-1}$ $\in S_{\tilde{p}}$, we see that $(N_K \beta)^{s^*}$ is in $S_{\tilde{p}}$ i.e. $N_K(\beta)$ is in $H_{\tilde{p}}$. Therefore, by the similar reasoning as in the proof of Theorem 1, Theorem 2 is proved. #

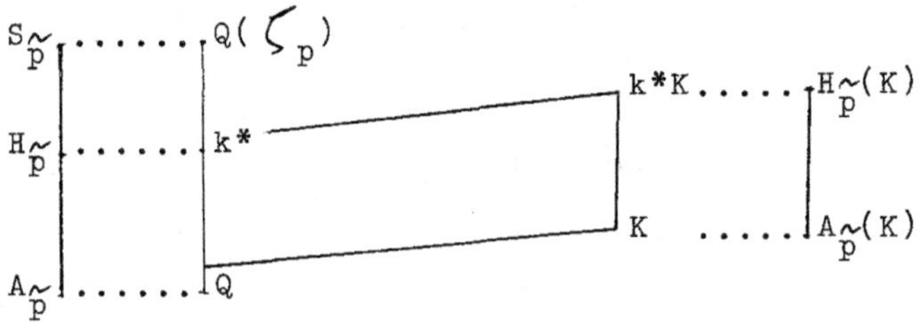

<u>Corollary</u>. In the situation of Theorem 2, if K is totally imaginary, then k^*K is unramified over K.

<u>Proof</u>. If K is totally imaginary, we have, for any number β $(\neq 0)$ in K, $N_K \beta > 0$. Hence, for any number β, prime to p, in K, we see that $N_K(\beta)^{s^*}$ is in $S_{\tilde{p}}$ i.e. $N_K(\beta)$ is in $H_{\tilde{p}}$. This implies that the ideal group, in K, corresponding to k^*K has the conductor 1. #

We also denote by $k^*(p)$ the absolute abelian number field k^* defined in Theorem 2 with respect to p and by $d^*(p)$ the degree $[k^*(p) : Q]$. Let $e(p)$ denote the greatest common divisor of the ramification indices of all the prime divisors, of p, in K. So we have

$$d^*(p) = (e(p), p - 1).$$

Then it follows that $k^*(p)$ is contained in $Q(\zeta_p)$ and

(3.7) $$d^*(p) \mid e(p) \quad \text{and} \quad e(p) \mid [K : Q] \; ;$$

p is totally ramified in $k^*(p)$ and all other rational prime numbers are unramified in $k^*(p)$. For the prime number 2, we put $k^*(2) = Q$ and $d^*(2) = 1$. Then the assertion of Theorem 2 holds for $p = 2$. Also for distinct rational prime numbers p_1, p_2, ... , p_t, let k_1^* be the composite $k^*(p_1)k^*(p_2)\ldots k^*(p_t)$. Then it is easily seen that k_1^*K is an abelian extension, which is unramified at all the finite prime ideals of K, and we have

$$[k_1^* : Q] = d^*(p_1)d^*(p_2)\ldots d^*(p_t)$$

(k_1^* is a subfield of $Q(\zeta_{p_1 p_2 \cdots p_t})$).

<u>Example 3</u>. 1) If $[K : Q]$ is even and, for a prime number $p \equiv 1 \pmod 4$, we have $(p) = \mathfrak{a}^2$ in K ($\mathfrak{a}$ is an ideal of K), then $K(\sqrt{p})$ is unramified over K. So if $\sqrt{p} \notin K$, we have

$2 \mid h_K$.

2) If $[K : Q]$ is divisible by an odd prime number ℓ and, for a prime number $p \equiv 1 \pmod{\ell}$, we have $(p) = \mathfrak{p}^{\ell}$ in K ($\mathfrak{p}$ is an ideal of K), then kK is unramified over K, where k is the unique subfield, of degree ℓ, of $Q(\zeta_p)_0$. So if $k \not\subset K$, we have $\ell \mid h_K$. In particular, if K is of Eisenstein type with respect to $p \equiv 1 \pmod{\ell}$, $[K : Q] = \ell$ and $K \not\subset Q(\zeta_p)$, then kK is unramified over K and so $\ell \mid h_K$. (An elementary proof of this last fact is given in Ishida [15]).

3) If $[K : Q]$ is even and, for a prime number $q \equiv 3 \pmod{4}$, we have $(q) = \mathfrak{L}^2$ in K ($\mathfrak{L}$ is an ideal of K), then $K(\sqrt{-q})$ is unramified at all the finite prime ideals of K. Moreover if K is totally imaginary, $K(\sqrt{-q})$ is unramified over K. So if $\sqrt{-q} \notin K$ (K is totally imaginary), we have $2 \mid h_K$. #

Chapter 4. The genus fields

Let K be an algebraic number field of finite degree n. As defined in Chapter 1, the genus field K^* of K is the maximal abelian extension of K, which is a composite of an absolute abelian number field k^* with K and is unramified at all the finite prime ideals of K. We may take k^* as the maximal absolute abelian subfield of K^* and do so in the following arguments. The extension degree of the genus field K^* over K is called the genus number g_K of K. We use these notations through the rest of this note.

Our purpose is to determine the absolute abelian number field k^* by the arithmetic properties of K.

Let $\tilde{f}$ be the conductor of k^* with the finite part f. So k^* is a subfield of the f-th cyclotomic number field $Q(\zeta_f)$. Take a prime divisor p of f. Then, as p is ramified in k^* and any prime divisors of p in K is unramified in k^*K, p must be ramified in K i.e. p must divide the discriminant D_K of K. That is, the ramification index (>1) of p in k^* divides that of any prime divisor of p in K and so divides the degree $[K : Q]$. More precisely, suppose that we have

(4.1) $p^t \| f$ and $(p) = \mathfrak{p}_1^{e_1} \mathfrak{p}_2^{e_2} \cdots \mathfrak{p}_m^{e_m}$ in K,

where $t > 0$ and $e_i > 0$. Then the ramification index $e^*(p)$ of p in k^* divides that of p in $Q(\zeta_f)$, which is known to be equal to $\varphi(p^t) = p^{t-1}(p - 1)$. On the other hand, as $\mathfrak{p}_1$, $\mathfrak{p}_2, \ldots, \mathfrak{p}_m$ are unramified in k^*K, $e^*(p)$ must divide e_1, $e_2, \ldots, e_m$ i.e. $e^*(p) \mid e = (e_1, e_2, \ldots, e_m)$. Hence if p does not divide at least one of e_i i.e. (p) is not equal to the p-th power $\mathfrak{a}^p$ of any ideal $\mathfrak{a}$ in K, then $e^*(p)$ is a divisor of $p - 1$ and so a divisor of $d^*(p) = (e, p - 1)$; consequently

(4.2) $e^*(p) \leqq d^*(p)$.

Keep the assumption that $(p) \neq \mathfrak{a}^p$ in K and let $k^*(p)$ be the unique subfield, of degree $d^*(p)$, of $Q(\zeta_p)$. By Theorem 2, $k^*(p)$ gives an abelian extension $k^*(p)K$ of K, which is unramified at all the finite prime ideals of K. Then, by definition, it follows that

(4.3) $k^*(p) \subset k^*$.

Since p is totally ramified in $k^*(p)$, we have

$$d^*(p) = (e, p - 1)$$
$$= \text{the ramification index of } p \text{ in } k^*(p).$$

Then (4.3) implies

(4.4) $$d^*(p) \leqq e^*(p).$$

Hence, by (4.2) and (4.3), we have

(4.5) $$e^*(p) = d^*(p)$$

and so we see that the unique prime divisor of p in $k^*(p)$ is unramified in k^* over $k^*(p)$ (for p such that $(p) \neq \mathfrak{N}^p$ in K).

Now the absolute abelian number field $k^*(p)$ is defined for all the rational prime numbers p right after the proof of Theorem 2 and, by definition, we have $k^*(p) \subset k^*$. On the other hand, denote by $e(p)$ the greatest common divisor of the ramification indices of all the prime divisors, of p, in K. Then we have

$$d^*(p) = [k^*(p) : Q] = (e(p), p - 1).$$

Consider all the prime divisors p of f such that

(4.6) $$(p) \neq \mathfrak{a}^p \text{ in } K \quad \text{i.e.} \quad p \nmid e(p),$$

where $\mathfrak{a}$ is any ideal of K, and construct the composite field of $k^*(p)$ for all such p's. That is, let

(4.7) $$k_1^* = \text{the composite field of } k^*(p) \text{ for all } p \text{ such that } p \mid f \text{ and } p \nmid e(p)$$
$$= \prod_{p \mid f,\ p \nmid e(p)} k^*(p) \quad \text{(composite)}.$$

For a rational prime number p not dividing f, p is unramified in k^* and so we have $k^*(p) = Q$. Clearly if $p \nmid D_K$, then we must have $e(p) = 1$ and so $p \nmid f$. Hence k_1^* is also written as

(4.8) $$k_1^* = \prod_{p \mid D_K,\ p \nmid e(p)} k^*(p) \quad \text{(composite)}.$$

Moreover, as each $k^*(p)$ is a subfield of the p-th cyclotomic number field $Q(\zeta_p)$ respectively, we have

(4.9) $$[k_1^* : Q] = \prod_p {}^* [k^*(p) : Q]$$

$$= \prod_p{}^{*} d^*(p) = \prod_{p \mid D_K,\ p \nmid e(p)} d^*(p),$$

where the product $\prod_p{}^{*}$ ranges over all p such that $p \mid f$ and $p \nmid e(p)$. On the other hand, let

(4.10) k_2^* = the intersection of the inertia fields $T^*(p)$ of (any) prime divisors of p in k^* over Q for all p such that $p \mid f$ and $p \nmid e(p)$

$$= \bigcap_{p \mid f,\ p \nmid e(p)} T^*(p).$$

Then all the rational prime numbers p such that $p \mid f$ and $p \nmid e(p)$ are unramified in k_2^* ; while they are ramified in k_1^*. Hence we have

(4.11) $$k_1^* \cap k_2^* = Q.$$

Consequently, denoting by $\mathfrak{T}^*(p)$ the inertia group, with respect to k^* over Q, of (any) prime divisors of p, we have, by (4.11), (4.5) and (4.7)

$$[k_1^* : Q] = [k_1^* k_2^* : k_2^*] \leqq [k^* : k_2^*]$$

$$= \left| \prod_p{}^{*} \mathcal{Z}^*(p) \right| \leqq \prod_p{}^{*} \left| \mathcal{Z}^*(p) \right|$$

$$= \prod_p{}^{*} e^*(p) = \prod_p{}^{*} d^*(p) = [k_1^* : Q],$$

where, for subgroups $\mathcal{H}, \mathcal{H}_1, \mathcal{H}_2, \ldots, \mathcal{H}_s$ of a finite abelian group $\mathcal{G}$, $|\mathcal{H}|$ denotes the order of $\mathcal{H}$ and $\prod_{i=1}^{s} \mathcal{H}_i$ denotes the subgroup, of $\mathcal{G}$, generated by $\mathcal{H}_1, \mathcal{H}_2, \ldots, \mathcal{H}_s$. Hence we have

$$k^* = k_1^* k_2^* \quad \text{(composite)}.$$

Thus we have the following

Theorem 3 (Ishida [18]). Let K be an algebraic number field of degree n and let k^* be the maximal abelian subfield of the genus field K^* of K; so we have $K^* = k^*K$. We define two subfields k_1^* and k_2^*, of k^*, by (4.7) and (4.10). Then we have

$$(4.12) \qquad k^* = k_1^* k_2^* \quad \text{(composite)} \quad \text{and} \quad k_1^* \cap k_2^* = Q.$$

Note that a rational prime number p is ramified in k_1^* only if we have $p \mid D_K$ and $p \nmid e(p)$ and a rational prime number

q is ramified in $k_2{}^*$ only if we have $q \mid D_K$ and $q \mid e(q)$, where D_K is the discriminant of K.

Here we also note that $q \mid e(q)$ implies $e(q) > 1$ and so $q \mid D_K$. If there is no prime number q such that $q \mid e(q)$, then $k_2{}^*$ is unramified over Q and we have $k_2{}^* = Q$ and $k^* = k_1{}^*$. Moreover $e(q)$ is a divisor of the degree $n = [K : Q]$. Consequently $q \mid e(q)$ implies that q divides n and D_K ; so if $(n, D_K) = 1$, then $k_2{}^* = Q$ and $k^* = k_1{}^*$. That is, we have the following

Corollary. Assume that there is no prime number q such that $q \mid e(q)$. (If $(n, D_K) = 1$, this assumption is satisfied.) Then

$$k^* = \prod_{p \mid D_K} k^*(p) \qquad \text{(composite)}$$

gives the genus field $K^* = k^*K$ of K. The Galois group of k^* over Q is the direct product of cyclic groups of order $d^*(p) = (e(p), p - 1)$ for all $p \mid D_K$. So the genus number g_K of K is given by

$$(4.13) \qquad g_K = \prod_{p \mid D_K} d^*(p) \,/\, [K_o : Q],$$

where K_o is the maximal abelian subfield of K.

Example 4. Let $K = Q(\zeta_{q^s})$ be the q^s-th cyclotomic number field with a prime q. Then, by the definition, we can easily see that the genus field K^* of K coincides with K : $K^* = K$. In this case, the two subfields k_1^* and k_2^* of $k^* = K^* = K$ in Theorem 3 are as follows : As is well known, $n = [K : Q]$ is $q^{s-1}(q - 1)$ and D_K is a power of q. Accordingly, $e(p) > 1$ $\Longleftrightarrow$ $p = q$ and $e(q) = n$.

i) If $s = 1$, then $q \nmid e(q) = q - 1$ and so $k_2^* = Q$ and $k_1^* = k^* = K$ (also, because $d^*(q) = (q - 1, q - 1) = q - 1$ and $k_1^* = k^*(q)$ is the unique subfield, of degree $q - 1$, of $Q(\zeta_q)$ i.e. $k_1^* = Q(\zeta_q) = K$).

ii) If $s \geqq 2$, then $q \mid e(q) = q^{s-1}(q - 1)$ and so $k_1^* = Q$ and $k_2^* = k^* = K$. #

Here we apply Theorem 3 and its Corollary to quadratic case and show the classical theorem of the genus theory : Let $K = Q(\sqrt{m})$ be a quadratic number field, where m is a square-free integer. Since K is a Galois extension of Q, of degree 2, we have

$$(4.14) \qquad p \mid D_K \Longleftrightarrow e(p) > 1 \Longleftrightarrow e(p) = 2$$

for a rational prime number p. Hence, by (4.8), we have

$$k_1^* = \prod_{p \mid D_K,\ p \neq 2} k^*(p) \quad \text{(composite)}. \tag{4.15}$$

On the other hand, for an odd prime number p dividing D_K, $k^*(p)$ is the unique subfield, of degree $d^*(p) = 2$, of $Q(\zeta_p)$ and so

$$k^*(p) = \begin{cases} Q(\sqrt{p}\,), & \text{if } p \equiv 1 \pmod 4 \\ Q(\sqrt{-p}), & \text{if } p \equiv 3 \pmod 4 \end{cases} = Q(\sqrt{p^*}\,), \tag{4.16}$$

where $p^* = (-1)^{(p-1)/2}\, p$ is the prime (quadratic) discriminant with respect to p.

Let $p_1, p_2, \ldots, p_t$ be all the prime divisors of D_K. Then first, by Corollary to Theorem 3, (4.15) and (4.16), if D_K is odd i.e. $m \equiv 1 \pmod 4$, we have

$$\begin{aligned} k^* &= Q(\sqrt{p_1^*}, \sqrt{p_2^*}, \ldots, \sqrt{p_t^*}\,), \\ K^* &= K(\sqrt{p_1^*}, \sqrt{p_2^*}, \ldots, \sqrt{p_t^*}\,) \\ &\text{and} \quad g_K = 2^{t-1}. \end{aligned} \tag{4.17}$$

The formula on g_K follows from the fact $\sqrt{p_1^*}\,\sqrt{p_2^*} \cdots \sqrt{p_t^*} = \sqrt{D_K} \in K$. Secondly, in the case of even D_K i.e. (e.g.) $p_1 = 2$, we have to consider as follows : Since 2 is the only ramifying prime number in k_2^*, we see that k_2^* is a subfield of the 2-power-th cyclotomic number field and so 2 is totally ramified in k_2^*. On the other hand, as k_2^*K is unramified at all prime divisors of 2 in K, we have

$$k_2^* \neq Q \Longleftrightarrow [k_2^* : Q] > 1 \Longleftrightarrow [k_2^* : Q] = 2$$
$$\Longrightarrow 2 \mid e(2) \Longleftrightarrow 2 \mid D_K.$$

Hence k_2^* must be a subfield of $Q(\zeta_8)$, of degree 1 or 2, i.e. $k_2^* = Q, Q(\sqrt{-1}\,), Q(\sqrt{2}\,)$ or $Q(\sqrt{-2}\,)$ and $k_2^* \neq Q$ implies that 2 is ramified in K i.e. $m \equiv 3$ or 2 (mod 4). Denote by p_1^* the prime (quadratic) discriminant with respect to 2 in the prime decomposition of D_K i.e.

$$p_1^* = -4, \quad 8 \quad \text{or} \quad -8$$

according to $m \equiv 3 \pmod 4$, $2 \pmod 8$ or $-2 \pmod 8$

respectively. We can write

$$D_K = p_1^* \, d,$$

where d is also a (quadratic) discriminant (or 1) such

that $d \equiv 1 \pmod 4$ and so $(p_1^*, d) = 1$. Then we have

$$L = Q(\sqrt{p_1^*})K = Q(\sqrt{p_1^*}, \sqrt{D_K}) = Q(\sqrt{p_1^*}, \sqrt{d})$$

and, for any rational prime number q,

$$\begin{aligned} q \nmid D_K &\Longrightarrow q \nmid p_1^*,\ q \nmid d \\ &\Longrightarrow q \text{ is unramified in } L. \\ &\Longrightarrow \text{any prime divisor of } q \text{ in } K \text{ is unramified in } L, \\ q \mid D_K &\Longrightarrow q \text{ is ramified in } K \text{ and } q \nmid p_1^* \text{ or } q \nmid d \\ &\Longrightarrow \text{the ramification index of } q \text{ in } L \text{ is } 2 \\ &\Longrightarrow \text{any prime divisor of } q \text{ in } K \text{ is unramified in } L. \end{aligned}$$

Hence L is unramified at all the finite prime ideals of K, which implies $Q(\sqrt{p_1^*}) \subset k^*$ i.e.

(4.18) $$k_2^* = Q(\sqrt{p_1^*}).$$

So, similarly as in the case of odd D_K, we have, by Theorem 3, (4.16) and (4.18),

$$k^* = Q(\sqrt{p_1^*}, \sqrt{p_2^*}, \ldots, \sqrt{p_t^*}),$$

(4.17)' $$K^* = K(\sqrt{p_1^*}, \sqrt{p_2^*}, \ldots, \sqrt{p_t^*})$$

$$\text{and} \quad g_K = 2^{t-1}.$$

Thus the classical theorem of the genus theory of quadratic number field is proved (again) in our method.

Though the above proof of (4.18) is very simple, here we give another one by using the 'Verschiebungssatz'. Clearly $Q(\sqrt{-1})$, $Q(\sqrt{2})$ and $Q(\sqrt{-2})$ correspond to the ideal groups

$$H_1 = \left\{ (a) \in A_2 \;\middle|\; a \equiv 1, 5 \pmod{^\times \tilde{8}} \right\},$$

$$H_2 = \left\{ (a) \in A_2 \;\middle|\; a \equiv 1, 7 \pmod{^\times \tilde{8}} \right\} \quad \text{and}$$

$$H_3 = \left\{ (a) \in A_2 \;\middle|\; a \equiv 1, 3 \pmod{^\times \tilde{8}} \right\}$$

with defining modulus $\tilde{8} = 8p_\infty$ respectively, where A_2 is the group of all ideals, prime to 2, in Q. Hence the abelian extension $Q(\sqrt{-1})K = K(\sqrt{-1})$ of K corresponds to the ideal group

$$H_1(K) = \left\{ \mathfrak{a} \in A_2(K) \;\middle|\; N_K \mathfrak{a} \equiv 1, 5 \pmod{^\times \tilde{8}} \right\}$$

in K and so $K(\sqrt{-1})$ is unramified at all the finite prime ideals of K if and only if the norms of any totally positive numbers, prime to 2, of K are (multiplicatively) congruent to 1 or 5 $(\mathrm{mod}^{\times}\ \tilde{8})$. Similar results hold for $K(\sqrt{2})$ and $K(\sqrt{-2})$. Now, for a totally positive integer $\gamma = x + y\sqrt{m}$ in K ($x, y \in Z$), prime to 2, we have

$$N_K\gamma = x^2 - y^2 m = \begin{cases} 1,\ 5 \quad (\mathrm{mod}^{\times}\tilde{8}), \text{ if } m \equiv 3 \pmod 4, \\ 1,\ 7 \quad (\mathrm{mod}^{\times}\tilde{8}), \text{ if } m \equiv 2 \pmod 8, \\ 1,\ 3 \quad (\mathrm{mod}^{\times}\tilde{8}), \text{ if } m \equiv -2 \pmod 8. \end{cases}$$

For example, if $m \equiv 3 \pmod 4$, γ is not prime to 2 if and only if $x \equiv y \pmod 2$. As $x \equiv 0 \pmod 2$ implies $x^2 \equiv 0$ or $4 \pmod 8$ and $x \equiv 1 \pmod 2$ implies $x^2 \equiv 1 \pmod 8$, we see that if γ is prime to 2 then $N_K\gamma \equiv 1, 5\ (\mathrm{mod}^{\times}\tilde{8})$. Hence we have

$$k_2^* = \begin{cases} Q(\sqrt{-1}), \text{ if } m \equiv 3 \pmod 4 \\ Q(\sqrt{2}), \text{ if } m \equiv 2 \pmod 8 \\ Q(\sqrt{-2}), \text{ if } m \equiv -2 \pmod 8 \end{cases} = Q(\sqrt{p_1^*}).$$

We add several remarks on the genus fields.

Let K be an algebraic number field of prime power degree

q^s. Then the ramifying rational prime numbers p in the absolute abelian number field k_2^* defined by (4.10) are those such that

$$p \mid e(p) > 1 \quad \text{and} \quad e(p) \mid [K : Q] = q^s ;$$

consequently q is the only ramifying prime number in k_2^* and so k_2^* is contained in the q-power-th cyclotomic number field. Of course, each $k^*(p)$ with $p \nmid e(p)$ is contained in the p-th cyclotomic number field. Hence, in this case (the prime power degree case), k_1^* and consequently the absolute abelian number field $k^* = k_1^* k_2^*$ is composed of absolute abelian number fields, whose conductors have a prime power as the finite part (cf. the case of quadratic number fields).

On the other hand, we have the following

Proposition 2. Let K be an algebraic number field of degree $n = n_1 n_2$ with $(n_1, n_2) = 1$. Suppose that K is composed of two subfields K_1 and K_2 such that $[K_1 : Q] = n_1$ and $[K_2 : Q] = n_2$ and let K_i^* be the genus field of K_i $(i = 1, 2)$. Then we have, for the genus field $K^* = k^* K$ of K,

(4.19) $$k^* = k^{(1)*} k^{(2)*} \quad \text{and so} \quad K^* = K_1^* K_2^*$$
(composite),

where $k^{(i)*}$ is the maximal abelian subfield of K_i^*.

Proof. By the definition, $k^{(i)*}K$ is composed of an absolute abelian number field with K and is unramified at all the finite prime ideals of K ; so we have $k^{(i)*}K \subset K^*$ i.e. $k^{(i)*} \subset k^*$ and consequently

$$k^{(1)*} k^{(2)*} \subset k^*.$$

On the other hand, let k $(\neq Q)$ be an absolute abelian number field of prime power degree q^u, for which kK is unramified at all the finite prime ideals of K. Then there is at least one rational prime number p ramifying in k and the ramification index of p in k over Q must divide $[k : Q] = q^u$ and $[K : Q] = n_1 n_2$. Hence we may assume that $q \mid n_1$ and $q \nmid n_2$. If $kK_1 = K_1$ i.e. $k \subset K_1$, then k is of course contained in $k^{(1)*}$. If $kK_1 \neq K_1$, then $[kK_1 : K_1]$ is a divisor (> 1) of $[k : Q] = q^u$. Suppose that a prime divisor $\mathfrak{p}$ of a rational prime number p in K_1 is ramified in kK_1 with the ramification index e^*, which is a power (> 1) of q. As kK is unramified over K, $\mathfrak{p}$ must be ramified in K and e^* is a divisor of $[K : K_1] = [K_2 : Q] = n_2$, which is a contradiction. So kK_1 is unramified at all the finite prime ideals of K_1 and k is contained in $k^{(1)*}$. Since any absolute abelian number field is composed of absolute abelian number fields of prime power degree, we have

$$k^* \subset k^{(1)*} \, k^{(2)*} . \quad \#$$

Finally we consider the case where K is an absolute abelian number field. As the genus field K^* of K is composed of an absolute abelian number field and K, K^* is also an absolute abelian number field and so we have

$$K^* = k^*.$$

Now, we prove the theorem of Leopoldt by using our method.

Theorem 4. Let K be an absolute abelian number field of finite degree n.

1) Let $n = q_1^{s_1} q_2^{s_2} \cdots q_t^{s_t}$, where q_i are distinct rational prime numbers $(s_i > 0)$. Then K is composed of absolute abelian number fields $K_1, K_2, \ldots, K_t$ with $[K_i : Q] = q_i^{s_i}$. Hence we have

$$(4.20) \qquad K^* = K_1^* \, K_2^* \cdots K_t^* \qquad \text{(composite)}.$$

2) Suppose $[K : Q] = q^s$ with a rational prime q $(s > 0)$.

Denoting by $e(p) = q^{e_p}$ the ramification index of a rational prime number p in K, we have

$$(4.21) \qquad K^* = \Big(\prod_{p \mid D_K,\ p \neq q} K^*(p)\Big)K$$

$$= \Big(\prod_{p \mid D_K,\ p \neq q} K^*(p)\Big)K^*(q) \qquad \text{(composite)},$$

where

$$(4.22) \qquad \begin{cases} K^*(p) = \text{the unique subfield, of degree } q^{e_p}, \text{ of } Q(\zeta_p), \text{ for } p \neq q. \\ K^*(q) = \text{a subfield, of degree } q^{e_q}, \text{ of some } q\text{-power-th cyclotomic number field.} \end{cases}$$

Note that if q is odd, then $K^*(q)$ is uniquely determined by its degree.

In particular, the genus field $K^* = k^*$ of an absolute abelian number field K is composed of absolute abelian number fields, whose conductors have a prime power as the finite part.

Proof. By the structure theorem of abelian groups and Proposition 2, we have 1) trivially. As for 2), we have, by (4.8),

$$k_1^* = \prod_{p \mid D_K,\ p \nmid e(p)} k^*(p) \qquad \text{(composite)},$$

where $k^*(p)$ is the subfield, of degree $d^*(p) = (e(p), p - 1)$, of $Q(\zeta_p)$. As

$$p \mid D_K \iff e(p) > 1$$
$$\iff e(p) \text{ is a power of } q > 1,$$

we have, for $p \mid D_K$, $p \neq q \iff p \nmid e(p)$. Moreover it is known that $e(p)$ is a divisor of $p^r(p - 1)$ with some $r \in Z$. Hence $p \nmid e(p)$ i.e. $p \neq q$ implies $e(p) \mid (p - 1)$ and so we have

$$d^*(p) = (e(p), p - 1) = e(p).$$

Accordingly, in the notations of Theorem, we have

$$k_1^* = \prod_{p \mid D_K,\ p \neq q} K^*(p) \qquad \text{(composite)}.$$

On the other hand, as remarked before Proposition 2, k_2^* defined by (4.10) is a subfield of some q-power-th cyclotomic number field and so q is totally ramified in k_2^*. Denote $q^u = [k_2^* : Q]$; we have $[k_2^* : Q] \mid q^{e_q}$ and so $u \leqq e_q$. Since q is not ramified in k_1^*, the inertia field T of q in K

over Q contains $k_1^* \cap K$ and therefore

$$[k_1^* \cap K : Q] \leq [T : Q] \quad \text{and} \quad [K : T] = q^{e_\ell}.$$

Now, as $k_1^*K \subset K^*$, $K^* = k^* = k_1^*k_2^*$ and $k_1^* \cap k_2^* = Q$ (cf. Theorem 3), we have

$$\begin{aligned}[k_1^* : Q] &= [k_1^*K : K]\,[k_1^* \cap K : Q] \\ &\leq [K^* : K]\,[T : Q] \\ &= [k_1^* : Q]\,[k_2^* : Q]\,[T : Q] \,/\, [K : Q] \\ &= [k_1^* : Q]\, q^u \,/\, [K : T] \\ &= [k_1^* : Q]\, q^{u-e_\ell} \leq [k_1^* : Q].\end{aligned}$$

Hence we have

$$[k_1^*K : K] = [K^* : K] \quad \text{and} \quad q^u = [k_2^* : Q] = q^{e_\ell};$$

and so we have

$$K^* = k_1^*K = \Big(\prod_{p \mid D_K,\; p \neq q} K^*(p) \Big)\, K$$

and k_2^* is a subfield, of degree q^{e_ℓ}, of $Q(\zeta_{q^\nu})$ with some $\nu \in Z$. #

<u>Corollary</u> (Leopoldt [23]). For an absolute abelian number field K, the genus number g_K is given by

$$g_K = \prod_p e(p) / [K : Q]. \tag{4.23}$$

<u>Proof</u>. If K has the prime power degree q^s, (4.23) follows from (4.21) and (4.22). Note that then g_K is a power of q. Now suppose that K is of degree $n = q_1^{s_1} q_2^{s_2} \dots q_t^{s_t}$ and $K = K_1 K_2 \dots K_t$ with $[K_i : Q] = q_i^{s_i}$. We prove (4.23) by the induction on t. Put $L = K_1 K_2 \dots K_{t-1}$. Accordingly we have $K = LK_t$ and $([L : Q], [K_t : Q]) = 1$. Then, assuming (4.23) for L and K_t, we see that the prime divisors of $g_L = [L^* : L]$ are in $\{ q_1, q_2, \dots, q_{t-1} \}$ and $g_K = [K_t^* : K_t]$ is a power of q_t and so we have

$$L^*K \cap K_t^*K = K, \quad L^* \cap K = L \quad \text{and} \quad K_t^* \cap K = K_t.$$

By Proposition 2, $K^* = L^*K_t^* = (L^*K)(K_t^*K)$ and so we have

$$\begin{aligned} g_K = [K^* : K] &= [L^*K : K]\,[K_t^*K : K] \\ &= [L^* : L]\,[K_t^* : K_t] = g_L\, g_{K_t}. \end{aligned}$$

Then (4.23) follows from the multiplicative property of degrees and ramification indices. #

<u>Example 5</u>. Let $K = Q(\zeta_m)$ be the m-th cyclotomic number field, which has the degree $[K : Q] = \varphi(m)$. On the other hand, for a rational prime number p, $p \mid D_K$ if and only if $p \mid m$ and we have $e(p) = \varphi(p^s)$ when $p^s \| m$. Let $m = \prod_{i=1}^{t} q_i^{s_i}$. Then we have, by (4.23),

$$g_K = \prod_{i=1}^{t} \varphi(q_i^{s_i}) \,/\, \varphi(m) = 1,$$

which shows $K^* = K$ (cf. Example 4). #

Chapter 5. The case of odd prime degree

Let K be an algebraic number field of odd prime degree q. We keep the notations in Chapter 4 : K^* is the genus field of K, $g_K = [K^* : K]$ is the genus number of K and k^* is the maximal absolute abelian subfield of K^*. Moreover let k_1^* and k_2^* be the subfields of k^* defined by (4.7) and (4.10) respectively. Then, by Theorem 3, we have

$$(5.1) \qquad k^* = k_1^* k_2^* \quad \text{(composite)} \quad \text{and} \quad k_1^* \cap k_2^* = Q.$$

For a rational prime number p, suppose that

$$(p) = \mathfrak{p}_1^{e_1} \mathfrak{p}_2^{e_2} \cdots \mathfrak{p}_m^{e_m} \quad \text{in } K.$$

Then, denoting $(e_1, e_2, \ldots, e_m)$ by $e(p)$, we see that $e(p)$ is a divisor of $q = [K : Q]$ and so

$$\begin{aligned} e(p) > 1 &\iff e(p) = q \\ &\iff p \text{ is totally ramified in } K, \\ p \mid e(p) &\iff p = q \text{ and } q \text{ is totally ramified in } K. \end{aligned}$$

Moreover

$$d^*(p) = (e(p),\ p-1) > 1$$
$$\Longleftrightarrow \quad e(p) = q \quad \text{and} \quad (q,\ p-1) > 1$$
$$\Longleftrightarrow \quad e(p) = q \quad \text{and} \quad p \equiv 1 \pmod q.$$

Let $p_1, p_2, \dots, p_t$ be all the rational prime numbers which satisfy the conditions :

$$(5.2) \qquad \begin{cases} \text{i)} \ \ p_i \ \text{ is totally ramified in } K, \\ \text{ii)} \ \ p_i \equiv 1 \pmod q. \end{cases}$$

Then we have $d^*(p_i) = q$ and so

$$(5.3) \qquad k_1^* = \prod_{i=1}^{t} k^*(p_i) \qquad \text{(composite)},$$

where $k^*(p_i)$ is the unique subfield, of degree q, of the p_i-th cyclotomic number field $Q(\zeta_{p_i})$. As for the subfield k_2^*, a rational prime number p is ramified in k_2^* if and only if $p \mid e(p)$. Hence, by the remark stated above, if q is not totally ramified in K, we have $k_2^* = Q$ and $k^* = k_1^*$ and therefore the genus field $K^* = k^*K$ of K is explicitly determined. On the other hand, suppose that q is totally ramified in K. Then, as q is the only ramifying prime in k_2^*, k_2^* is a subfield of $Q(\zeta_{q^s})$ for some $s \in Z$ (cf. Chapter 4) and so q is totally ramified in k_2^*, which implies that $[k_2^* : Q]$ is a divisor of $q = [K : Q]$ i.e. $[k_2^* : Q] = 1$ or q i.e.

(5.4) $k_2^* = Q$ or the unique subfield k_o, of degree q, of $Q(\zeta_{q^2})$.

Hence our problem is to obtain a necessary and sufficient condition for $k_2^* = k_o$ i.e. k_oK to be unramified over K under the assumption that q is totally ramified in K. Note that, as $[k_o : Q] = q$ is odd, there is no ramification of infinite prime divisors, in k_oK, of K.

As in Chapter 3, we use terminologies of class field theory. Clearly k_o is contained in the maximal real subfield $Q(\zeta_{q^2})_o$ of $Q(\zeta_{q^2})$ and corresponds to the ideal group

$$H_{q^2} = \left\{ (a) \in A_q \;\middle|\; a^{q-1} \equiv 1 \pmod{^\times q^2} \right\},$$

in Q, with defining modulus q^2, where A_q is the group of all ideals, prime to q, in Q. We have

$$(A_q : H_{q^2}) = q.$$

Then the 'Verschiebungssatz' implies that k_oK is the abelian extension of K corresponding to the ideal group

$$H_{q^2}(K) = \left\{ \mathfrak{a} \in A_q(K) \;\middle|\; N_K \mathfrak{a}^{q-1} \equiv 1 \pmod{^\times q^2} \right\},$$

in K, with defining modulus q^2, where $A_q(K)$ is the group of all the ideals, prime to q, in K. Hence we see that k_oK is unramified over K if and only if $H_{q^2}(K)$ contains all the principal ideals, prime to q, in K i.e. for any integer γ, prime to q, in K, $N_K(\gamma)^{q-1} \equiv 1 \pmod{q^2}$ (cf. the proofs of Theorems 1 and 2).

Since q is assumed to be totally ramified in K, we can find, by Proposition 1, a primitive element π of K, whose minimal polynomial is of Eisenstein type with respect to q :

$$f(X) = X^q + a_1X^{q-1} + \dots + a_{q-1}X + a_q \in Z[X] \tag{5.5}$$

with $q \mid a_i$ $(i = 1, 2, \dots, q-1)$ and $q \parallel a_q$.

First suppose that k_oK is unramified over K. Then, for the integers $\gamma = 1 - y\pi$ in K $(y \in Z)$, we must have

$$N_K \gamma^{q-1} = N_K(\gamma)^{q-1} \equiv 1 \pmod{^\times q^2}.$$

On the other hand, as $N_K \gamma = y^q f(1/y)$ $(y \neq 0)$ and $q \mid a_i$, we have

$$N_K \gamma^{q-1} = (1 + a_1 y + \dots + a_q y^q)^{q-1}$$
$$\equiv 1 + (q - 1)(a_1 y + \dots a_q y^q)$$
$$\equiv 1 - (a_1 y + \dots a_q y^q) \qquad (\mathrm{mod}\ q^2).$$

So, by writing $a_i = qb_i$ $(b_i \in Z)$, we have

$$a_1 y + a_2 y^2 + \dots + a_q y^q = qy(b_1 + b_2 y + \dots + b_q y^{q-1})$$
$$\equiv 0 \quad (\mathrm{mod}\ q^2)$$
$$\text{i.e.} \quad b_1 + b_2 y + \dots + b_q y^{q-1} \equiv 0 \quad (\mathrm{mod}\ q)$$

for $y = 1, 2, \dots, q - 1$, which implies that, as $q \| a_q$ and so $q \nmid b_q$,

$$b_1 + b_2 Y + \dots + b_q Y^{q-1} \equiv b_q (Y^{q-1} - 1) \quad (\mathrm{mod}\ q)$$

as a polynomial of Y over Z. Hence there must hold the congruence relation

$$b_2 \equiv \dots \equiv b_{q-1} \equiv b_1 + b_q \equiv 0 \quad (\mathrm{mod}\ q),$$

i.e. the coefficients a_i of $f(X)$ satisfy the following conditions :

(5.6) $$a_2 \equiv \ldots \equiv a_{q-1} \equiv a_1 + a_q \equiv 0 \pmod{q^2}.$$

Note that, as $q \| a_q$, (5.6) implies $q \| a_1$ also.

Conversely suppose that the coefficients a_i of the minimal polynomial $f(X)$ of π satisfy the conditions (5.6). We need the following

<u>Lemma 5</u>. Let $X_1, X_2, \ldots, X_q$ be independent variables and consider a monomial $M = X_1^{k_1} X_2^{k_2} \ldots X_q^{k_q}$ with $k_1 \geqq 2$. Let $F(X_1, X_2, \ldots, X_q)$ be the 'smallest' symmetric polynomial containing M as its term. Using the fundamental symmetric polynomials $Y_1 = X_1 + X_2 + \ldots + X_q$, $Y_2 = X_1X_2 + \ldots + X_iX_j + \ldots + X_{q-1}X_q$, ..., $Y_q = X_1X_2 \ldots X_q$, we can write

(5.7) $$F(X_1, X_2, \ldots, X_q) = c + aY_1 + bY_q + \ldots \in Z[Y_1, Y_2, \ldots, Y_q].$$

Then we have $c = a = 0$ and $b \equiv 0 \pmod q$.

<u>Proof</u>. Trivially we have

$$c = F(0, 0, \ldots, 0) = 0,$$
$$a = \partial F/\partial X_1 (0, 0, \ldots, 0) = 0.$$

On the other hand, consider the coefficients s_N of $X_1X_2 \ldots X_q$

in a monomial $N = Y_1^{h_1} Y_2^{h_2} \dots Y_q^{h_q}$ as a polynomial of $X_1, X_2, \dots, X_q$. Of course, we may restrict our consideration to such an N with $h_1 + 2h_2 + \dots + qh_q = q$. Then

$$\text{i)} \quad h_q \neq 0 \implies N = Y_q \implies s_N = 1,$$

$$\begin{aligned} \text{ii)} \quad h_q = 0 &\implies \text{for an index } j \ (< q),\ h_j \neq 0 \\ &\implies N = Y_1^{h_1} \dots \left(\sum X_{i_1} \dots X_{i_j}\right)^{h_j} \dots Y_q^{h_q} \\ &\implies s_N \text{ is a multiple of } {}_qC_j \\ &\implies s_N \equiv 0 \pmod{q}. \end{aligned}$$

Hence, comparing the coefficients of $X_1 X_2 \dots X_q$ in (5.7), we have

$$\begin{aligned} 0 &= \partial^q F / \partial X_1 \partial X_2 \dots \partial X_q (0, 0, \dots, 0) \\ &\equiv b \pmod{q}. \quad \# \end{aligned}$$

For example, let $q = 3$, $M = X_1^2 X_2$. Then we have

$$\begin{aligned} &F(X_1, X_2, X_3) \\ &= X_1^2 X_2 + X_1^2 X_3 + X_1 X_2^2 + X_2^2 X_3 + X_1 X_3^2 + X_2 X_3^2 \\ &= (X_1 + X_2 + X_3)(X_1 X_2 + X_1 X_3 + X_2 X_3) - 3X_1 X_2 X_3 \\ &= Y_1 Y_2 - 3Y_3. \end{aligned}$$

<u>Corollary</u>. In our case (i.e. under the assumption (5.6)),

let $\pi = \pi^{(1)}, \pi^{(2)}, \ldots, \pi^{(q)}$ be all the conjugates of π over Q. Then we have

$$F(\pi^{(1)}, \pi^{(2)}, \ldots, \pi^{(q)}) \equiv 0 \pmod{q^2}.$$

Proof. As $q^2 \mid a_2, \ldots, q^2 \mid a_{q-1}$, we have

$$F(\pi^{(1)}, \pi^{(2)}, \ldots, \pi^{(q)})$$
$$\equiv b\pi^{(1)}\pi^{(2)} \ldots \pi^{(q)} = -ba_q \equiv 0 \pmod{q^2}. \quad \#$$

Now let $(q) = \mathfrak{q}^q$ in K, where $\mathfrak{q}$ is a prime ideal of K. Then we have $\mathfrak{q} \parallel \pi$ (cf. Chapter 2). Let Q_q be the q-adic completion of Q and $K_{\mathfrak{q}}$ the $\mathfrak{q}$-adic completion of K. As $K_{\mathfrak{q}}$ is totally ramified over Q_q, π is a primitive element of K and $1, \pi, \ldots, \pi^{q-1}$ constitute the integral basis of $K_{\mathfrak{q}}$ over Q_q (cf. Chapter 2). So any integer Γ in $K_{\mathfrak{q}}$ can be written as

$$\Gamma = x_o + x_1\pi + \ldots + x_{q-1}\pi^{q-1}$$

with integers x_i in Q_q. Then, by Corollary to Lemma 5 and $q^2 \mid a_2, \ldots, q^2 \mid a_{q-1}$, we have

$$(5.8) \qquad N_{K_{\mathfrak{q}}/Q_q}\Gamma = N_{K_{\mathfrak{q}}/Q_q}(x_o + x_1\pi + \ldots + x_{q-1}\pi^{q-1})$$

$$\equiv N_{K_{\mathfrak{q}}/Q_q}(x_0 + x_1\pi)$$
$$= x_0^{\,q} - a_1 x_0^{\,q-1} x_1 + \dots - a_q x_1^{\,q}$$
$$\equiv x_0^{\,q} - a_1 x_0^{\,q-1} x_1 - a_q x_1^{\,q}$$
$$= x_0^{\,q} - x_1(a_1 x_0^{\,q-1} + a_q x_1^{\,q-1}) \pmod{q^2}.$$

Then Γ is prime to $\mathfrak{q}$ if and only if x_0 is prime to q. Moreover we have, in (5.8),

$$q \mid x_1 \implies x_1(a_1 x_0^{\,q-1} + a_q x_1^{\,q-1}) \equiv 0 \pmod{q^2},$$
$$q \nmid x_1 \quad (q \nmid x_0) \implies x_1(a_1 x_0^{\,q-1} + a_q x_1^{\,q-1})$$
$$\equiv x_1(a_1 + a_q) \equiv 0 \pmod{q^2}$$

Hence, for any integer Γ, prime to $\mathfrak{q}$, in $K_{\mathfrak{q}}$, we have, by (5.8),

$$N_{K_{\mathfrak{q}}/Q_q}\Gamma \equiv x_0^{\,q} \pmod{q^2}$$
$$\text{and so} \quad N_{K_{\mathfrak{q}}/Q_q}\Gamma^{q-1} \equiv 1 \pmod{q^2}.$$

Accordingly, for any integer γ, prime to $\mathfrak{q}$, in K, we have

$$N_K(\gamma)^{q-1} = N_K\gamma^{q-1} = N_{K_{\mathfrak{q}}/Q_q}\gamma^{q-1} \equiv 1 \pmod{q^2},$$

which implies, as remarked above, k_0K is unramified over K.

Though the case where K is abelian (i.e. cyclic) over Q is already treated in Chapter 4, we reconsider it here. We need the following

Lemma 6. Let L be an absolute abelian number field, in which an odd prime number q is totally ramified. Then L is cyclic over Q.

Proof. L is contained in a cyclotomic number field : $L \subset Q(\zeta_m)$. Writing $m = q^s m_0$ with $q \nmid m_0$, we have

$$L \cap Q(\zeta_{m_0}) = Q \quad \text{and} \quad Q(\zeta_{q^s}) \cap Q(\zeta_{m_0}) = Q.$$

On the other hand, as q is odd, $Q(\zeta_{q^s})$ is cyclic over Q and so $Q(\zeta_m) = Q(\zeta_{q^s})Q(\zeta_{m_0})$ is cyclic over $Q(\zeta_{m_0})$, which contains $LQ(\zeta_{m_0})$. Hence $LQ(\zeta_{m_0})$ is cyclic over $Q(\zeta_{m_0})$ and so L is cyclic over Q. #

Corollary. In the case where K is abelian over Q (of degree q), if q is totally ramified in K, then k_0K is unramified over K (i.e. (5.6) is satisfied).

Proof. Suppose that $k_0K \neq K$ and k_0K is ramified over K. Then k_0K is of degree q over K and so is totally ramified over K. That is, q is totally ramified in the absolute abelian number field k_0K and consequently, by Lemma 6,

k_oK is cyclic, of degree q^2, over Q. Hence k_oK has the unique subfield of degree q and so we must have $k_o = K$ i.e. $k_oK = K$, which is a contradiction. #

Summarizing the results just obtained, we have the following

<u>Theorem 5</u> (Ishida [19]). Let q be an odd prime number and let K be an algebraic number field of degree q. Let p_1, p_2, ... , p_t be all the rational numbers such that p_i is totally ramified in K and $p_i \equiv 1 \pmod{q}$. Then we have

$$k_1^* = \prod_{i=1}^{t} k^*(p_i) \qquad \text{(composite)},$$

where $k^*(p_i)$ is the unique subfield, of degree q, of $Q(\zeta_{p_i})$. Moreover when q is totally ramified in K, take a primitive element π of K, whose minimal polynomial $f(X) = X^q + a_1X^{q-1} + \dots + a_q \in Z[X]$ is of Eisenstein type with respect to q. Consider the conditions

$$(5.6) \qquad a_2 \equiv \dots \equiv a_{q-1} \equiv a_1 + a_q \equiv 0 \pmod{q^2}.$$

Then we have

$$k_2^* = \begin{cases} \text{the unique subfield } k_0, \text{ of degree } q, \text{ of} \\ \quad Q(\zeta_{q^2}), \text{ if } q \text{ is totally ramified in} \\ \quad K \text{ and (5.6) is satisfied,} \\ Q, \text{ otherwise.} \end{cases}$$

Thus, for the absolute abelian number field $k^* = k_1^* k_2^*$, $K^* = k^*K$ is the genus field of K. In particular, when K is cyclic over Q, any rational prime number p ($\neq q$) ramifying in K is totally ramified in K and satisfies $p \equiv 1 \pmod q$, and if q is totally ramified in K, (5.6) is satisfied. Hence the genus number g_K of K is given as follows :

i) K is not cyclic over Q.

$$(5.9) \qquad g_K = \begin{cases} q^{t+1}, \text{ if } q \text{ is totally ramified in } K \text{ and} \\ \quad (5.6) \text{ is satisfied,} \\ q^t, \text{ otherwise.} \end{cases}$$

ii) K is cyclic over Q.

$$(5.9) \qquad g_K = \begin{cases} q^t, \text{ if } q \text{ is totally ramified in } K, \\ q^{t-1}, \text{ otherwise.} \end{cases}$$

<u>Corollary</u>. For the q-rank $d = d^{(q)}C_K$ of the ideal class group C_K of K, we have

i) K is not cyclic over Q.

(5.10) $$d \geqq \begin{cases} t+1, & \text{if } q \text{ is totally ramified in } K \text{ and (5.6) is satisfied,} \\ t, & \text{otherwise.} \end{cases}$$

ii) K is cyclic over Q.

(5.10) $$d \geqq \begin{cases} t, & \text{if } q \text{ is totally ramified in } K, \\ t-1, & \text{otherwise.} \end{cases}$$

Now, in Theorem 5, we consider the congruence relations (5.6) and show that there is an algorithm to decide whether (5.6) is satisfied or not i.e. $k_o K$ is unramified over K or not.

That is, let K be an algebraic number field of odd prime degree q, in which q is totally ramified and let k_o be the unique subfield, of degree q, of the q^2-th cyclotomic number field $Q(\zeta_{q^2})$. If we can find a primitive element π of K (explicitly), whose minimal polynomial

$$f(X) = X^q + a_1 X^{q-1} + \ldots + a_q \in Z[X]$$

is of Eisenstein type with respect to q, then we see, by Theorem

5, that k_oK is unramified over K if and only if

$$(5.6) \qquad a_2 \equiv \dots \equiv a_{q-1} \equiv a_1 + a_q \equiv 0 \pmod{q^2}$$

is satisfied.

A typical example is as follows :

Example 6 (Fröhlich [7]). Let $K = Q(\sqrt[q]{a})$ be a pure number field of degree q, where a (>1) is a q-power free integer.

i) If $q^s \| a$ $(0<s<q)$, we can find integers $m, n \in Z$ $(1 \leq m \leq q-1, n>0)$ such that $ms = 1 + nq$ and so

$$K = Q(\sqrt[q]{a}^{\,m}/q^n) = Q(\sqrt[q]{b})$$

with $b \in Z$ and $q \| b$. Then the minimal polynomial $X^q - b$ of $\sqrt[q]{b}$ is of Eisenstein type with respect to q but (5.6) is not satisfied : $0 - b = -b \not\equiv 0 \pmod{q^2}$. Accordingly k_oK is not unramified over K.

ii) If $q \nmid a$, then, putting $\pi = \sqrt[q]{a} - a$, we have $K = Q(\pi)$ and

$$\pi^q + qa\pi^{q-1} + q(q-1)/2\cdot a^2\pi^{q-2} + \dots$$
$$+ qa^{q-1}\pi + (a^q - a) = 0.$$

It is known that if q is totally ramified in $K = Q(\sqrt[q]{a})$, then $q \parallel a^q - a$ and so the minimal polynomial of π is of Eisenstein type with respect to q. But (5.6) is not satisfied : $qa^{q-1} \not\equiv 0 \pmod{q^2}$. Accordingly k_0K is not unramified over K. #

Now, back to the general situation, let K be an algebraic number field of degree q, in which q is totally ramified : $(q) = \mathfrak{q}^q$, where $\mathfrak{q}$ is a prime ideal of K and $N_K\mathfrak{q} = q$. Let β be a primitive element of K, whose minimal polynomial is

(5.11) $$g(X) = X^q + b_1X^{q-1} + \dots + b_{q-1}X + b_q \in Z[X].$$

Then, as q is totally ramified in K, we have

$$g(X) \equiv (X - t)^q \pmod{q}$$

with $t \in Z$ and so it must hold

(5.12) $$b_1 \equiv b_2 \equiv \dots \equiv b_{q-1} \equiv 0 \pmod{q}$$

(cf. Chapter 2). Of course, (5.12) holds for the minimal polynomials of any primitive elements of K, which are integers.

Let $q^e \parallel b_q = -N_K \beta$ and

$$e = rq + s \quad \text{with} \quad r, s \in Z \quad \text{and} \quad 0 \leqq s < q.$$

Then, considering the prime ideal decomposition in K, we have $(q)^r = \mathfrak{q}^{rq} \mid \beta$ and so $\beta_1 = \beta / q^r$ is an integer of K, whose minimal polynomial is given by

$$g_1(X) = X^q + b_1/q^r \cdot X^{q-1} + \dots$$
$$+ b_{q-1}/q^{r(q-1)} \cdot X + b_q/q^{rq} \in Z[X].$$

Hence, as remarked above, the coefficients of $g_1(X)$ must satisfy the similar congruence relation as (5.12), that is,

(5.13) $$b_1 \equiv 0 \pmod{q^{r+1}}, \quad b_2 \equiv 0 \pmod{q^{2r+1}}, \dots, \quad b_{q-1} \equiv 0 \pmod{q^{(q-1)r+1}}.$$

When $q^{rq} \mid b_q$, these conditions are necessary ones for q to be totally ramified in K. Consequently, under (5.13), we replace β by β_1 if necessary and we may assume that, in (5.11),

(5.14) $$q^s \parallel b_q \Longrightarrow 0 \leqq s \leqq q - 1 \qquad (\mathfrak{q}^s \parallel \beta).$$

1) If $s = 1$ i.e. $q \| b_q$, then, by (5.12), $g(X)$ is of Eisenstein type with respect to q and so k_oK is unramified over K if and only if

$$b_2 \equiv \dots \equiv b_{q-1} \equiv b_1 + b_q \equiv 0 \pmod{q^2}.$$

2) If $1 < s \leqq q - 1$, we can find integers $m, n \in Z$ ($1 \leqq m \leqq q - 1$, $n > 0$) such that $ms = 1 + nq$. Then $\gamma = \beta^m/q^n$ is an integer of K, which is also a primitive element of K, and we have $q \| N_K\gamma$. Hence the minimal polynomial $h(X)$ of γ is of Eisenstein type with respect to q and so, by observing the coefficients of $h(X)$, we can tell whether k_oK is unramified over K or not.

Example 7. Let $q = 3$ and K a cubic number field, in which 3 is totally ramified. Suppose that

$$(5.15) \qquad g(X) = X^3 + b_1X^2 + b_2X + b_3 \in Z[X],$$
$$3 \mid b_1,\ 3 \mid b_2,\ 3^2 \| b_3$$

is the minimal polynomial of a primitive element β of K. As $2\cdot 2 = 1 + 1\cdot 3$, $\gamma = \beta^2/3$ is an integer of K and the minimal polynomial of γ is given by

$$h(X) = X^3 + c_1X^2 + c_2X + c_3 \in Z[X],$$

$$c_1 = -(b_1^2 - 2b_2)/3, \quad c_2 = (b_2^2 - 2b_1b_3)/3^2,$$

$$c_3 = -b_3^2/3^3 \quad ;$$

in particular, we have $3 \parallel c_3$. As $3 \mid c_1$ and $3 \mid c_2$, we have $3^2 \mid b_2$. So $h(X)$ is of Eisenstein type with respect to 3 and we see that k_oK is unramified over K if and only if $c_2 \equiv c_1 + c_3 \equiv 0 \pmod{3^2}$. Then

$$3^2 \mid c_2 \iff 3^4 \mid b_1b_3 \iff 3^2 \mid b_1$$
$$\text{and so} \quad 3 \parallel c_1 \iff 3^2 \parallel b_2$$

and, putting $b_2 = 3^2d_2$, $b_3 = 3^2d_3$ $(3 \nmid d_2, d_3)$, we have

$$3^2 \mid c_1 + c_3 \iff 2\cdot 3d_2 - 3d_3^2 \equiv 0 \pmod{3^2}$$
$$\iff 2d_2 - d_3^2 \equiv 0 \pmod{3} \iff 2d_2 \equiv 1 \pmod{3}$$
$$\iff d_2 \equiv 2 \pmod{3}.$$

Hence, for a cubic number field $K = Q(\beta)$, where β is a root of (5.15), we see that

$$k_oK \text{ is unramified over } K$$
$$\iff 3^2 \mid b_1,\ 3^2 \parallel b_2 \quad \text{and} \quad b_2/3^2 \equiv 2 \pmod{3}.$$

From the conditions of the right side or from even weaker conditions $3 \mid b_1$, $3^2 \mid b_2$ $(3^2 \| b_3)$, it follows that $h(X)$ is of Eisenstein type with respect to 3 and is irreducible in $Q[X]$ and so 3 is totally ramified in $K = Q(\beta)$. #

Example 8. Let $q = 5$ and K an algebraic number field of degree 5, in which 5 is totally ramified. Suppose that

$$(5.16) \qquad g(X) = X^5 + b_1X^4 + b_2X^3 + b_3X^2 + b_4X + b_5 \in Z[X],$$
$$5 \mid b_1, \ldots , 5 \mid b_4, 5^3 \| b_5$$

is the minimal polynomial of a primitive element β of K. As $2 \cdot 3 = 1 + 1 \cdot 5$, $\gamma = \beta^2/5$ is an integer of K and the minimal polynomial of γ is given by

$$h(X) = X^5 + c_1X^4 + c_2X^3 + c_3X^2 + c_4X + c_5 \in Z[X],$$
$$c_1 = -(b_1{}^2 - 2b_2)/5,$$
$$c_2 = (b_2{}^2 - 2b_1b_3 + 2b_4)/5^2,$$
$$c_3 = -(b_3{}^2 - 2b_2b_4 + 2b_1b_5)/5^3,$$
$$c_4 = (b_4{}^2 - 2b_3b_5)/5^4, \quad c_5 = -b_5{}^2/5^5 \ ;$$

in particular, we have $5 \| c_5$. As $5 \mid c_1, \ldots , 5 \mid c_4$, we have

$5^2 \mid b_2$, $5^3 \mid b_4$, $5^2 \mid b_3$. So $h(X)$ is of Eisenstein type with respect to 5 and we see that k_oK is unramified over K if and only if $c_2 \equiv c_3 \equiv c_4 \equiv c_1 + c_5 \equiv 0 \pmod{5^2}$. Then

$$5^2 \mid c_4 \iff 5^6 \mid b_3 b_5 \iff 5^3 \mid b_3$$
$$\text{and so } 5^2 \mid c_3 \iff 5^5 \mid b_1 b_5 \iff 5^2 \mid b_1$$
$$\text{and so } 5^2 \mid c_2 \iff 5^4 \mid b_4$$
$$\text{and so } 5 \parallel c_1 \iff 5^2 \parallel b_2$$

and, putting $b_2 = 5^2 d_2$, $b_5 = 5^3 d_5$ $(5 \nmid d_2, d_5)$, we have

$$5^2 \mid c_1 + c_5 \iff 2 \cdot 5d_2 - 5d_5{}^2 \equiv 0 \pmod{5^2}$$
$$\iff 2d_2 - d_5{}^2 \equiv 0 \pmod{5}$$
$$\iff 2d_2 \equiv d_5{}^2 \pmod{5}$$
$$\iff d_2 \equiv 3,\ d_5 \equiv \pm 1 \pmod{5}$$
$$\text{or} \quad d_2 \equiv 2,\ d_5 \equiv \pm 2 \pmod{5}.$$

Hence, for an algebraic number field $K = Q(\beta)$ of degree 5, where β is a root of (5.16), we see that

$$k_oK \text{ is unramified over } K$$
$$\iff 5^2 \mid b_1,\ 5^2 \parallel b_2,\ 5^3 \mid b_3,\ 5^4 \mid b_4 \text{ and}$$
$$b_2/5^2 \equiv 3,\ b_5/5^3 \equiv \pm 1 \pmod{5}$$

$$\text{or} \quad b_2/5^2 \equiv 2,\ b_5/5^3 \equiv \pm 2 \pmod 5.$$

From the conditions of the right side or from even weaker conditions $5 \mid b_1$, $5^2 \mid b_2$, $5^3 \mid b_4$, $5^2 \mid b_3$ $(5^3 \parallel b_5)$, it follows that $h(X)$ is of Eisenstein type with respect to 5 and is irreducible in $Q[X]$ and so 5 is totally ramified in $K = Q(\beta)$. #

3) If $s = 0$ i.e. $q \nmid b_q$, we take $\beta' = \beta + b_q$ in place of β, whose minimal polynomial is given by

$$g(X - b_q) = X^q + \dots + g(-b_q) \in Z[X],$$

$$g(-b_q) = (b_q - b_q{}^q) + b_1 b_q{}^{q-1} - \dots$$
$$- b_{q-1} b_q \equiv 0 \pmod q.$$

Accordingly, as remarked above, there is an $r_1 \in Z$ $(r_1 \geqq 0)$ such that $\beta_1 = \beta'/q^{r_1}$ is an integer of K, whose minimal polynomial

$$g_1(X) = X^q + \dots + g(-b_q)/q^{r_1 q} \in Z[X]$$

satisfies the condition $q^{s_1} \parallel g(-b_q)/q^{r_1 q}$ with $0 \leqq s_1 \leqq q - 1$. Hence if $1 \leqq s_1 \leqq q - 1$, we can reduce our case 3) to 1) or 2) just considered above. If $s_1 = 0$, we continue the above process. Note if $s_1 = 0$, then, as $q \mid g(-b_q)$, we have $r_1 > 0$. However these processes can not be continued in infinite steps. In fact, otherwise, there are, for any natural number j, integers $b_q, b_q', \dots, b_q^{(j)} \in Z$, $r_1, r_2, \dots, r_j \in Z$ $(r_i > 0)$ such that

$$\cfrac{\cfrac{\cfrac{\beta + b_q}{q^{r_1}} + b_q'}{q^{r_2}} + b_q''}{q^{r_3}} + \dots$$

is also an integer of K. This implies that there is an integer, of K, of the form

$$(\beta + c_j)/q^{r_1 + r_2 + \dots + r_j}$$

with $c_j \in Z$. Then we have $q^{r_1 + r_2 + \dots + r_j} \mid (O_K : Z[\beta])$ and so $q^j \mid (O_K : Z[\beta])$, which is a contradiction. Hence the case 3) is always reduced to the case 1) or 2).

Thus we obtain an algorithm to decide whether $k_o K$ is unramified over K or not.

Chapter 6. The genus fields of cubic number fields

In this chapter, we apply the results of Chapters 4 and 5 to a cubic number field K and determine the genus field K^* of K explicitly.

Let $K = Q(\alpha)$ be a cubic number field and let the minimal polynomial $f(X)$ of a primitive element α of K be of the following form

$$(6.1) \qquad f(X) = X^3 + aX + b \in Z[X].$$

This last condition does not affect the generality. Moreover we may assume that there is no rational prime number p such that

$$(6.2) \qquad p^2 \mid a \quad \text{and} \quad p^3 \mid b.$$

Then we have, for the discriminants D_f of $f(X)$ and D_K of K,

$$(6.3) \qquad D_f = -(4a^3 + 27b^2) = (O_K : Z[\alpha])^2 D_K.$$

As is shown in Chapter 5, we can construct the genus field K^* of K explicitly, if we know all the rational prime numbers p such that p is totally ramified in K and $p \equiv 1 \pmod 3$ and if we could decide whether k_oK is unramified over K or not, where k_o is the unique cubic subfield of $Q(\zeta_9)$.

<u>Lemma 7</u>. For a rational prime number $p \neq 3$, p is totally ramified in K if and only if

$$(6.4) \qquad p \mid a,\ p \parallel b \quad \text{or} \quad p^2 \mid a,\ p^2 \parallel b.$$

<u>Proof</u>. If p is totally ramified in K, we have

$$X^3 + aX + b \equiv (X - t)^3 \pmod p$$

(cf. Takagi [26]), which implies that $p \mid t$ and so $p \mid a$ and $p \mid b$. Conversely suppose $p \mid (a, b)$. Noting (6.2), we consider the following four cases :

1) If $p \mid a$, $p \parallel b$, then $f(X)$ is of Eisenstein type with respect to p and so p is totally ramified in K.

2) If $p \parallel a$, $p^2 \parallel b$ and p is totally ramified in K, then, similarly as in Example 7, $\gamma = \alpha^2/p$ is an integer of K and the minimal polynomial of γ is given by

$$h(X) = X^3 + 2a/p \cdot X^2 + a^2/p^2 \cdot X - b^2/p^3 \in Z[X],$$
$$p \parallel b^2/p^3.$$

As p is totally ramified in K, we have

$$h(X) \equiv (X - r)^3 \pmod{p},$$

which implies that $p \mid r^3$ and so $p \mid a^2/p^2$. Accordingly we have a contradiction.

3) If $p^2 \mid a$, $p^2 \parallel b$, then the same reasoning as in 2) shows that the minimal polynomial $h(X)$ of $\gamma = \alpha^2/p$ is of Eisenstein type with respect to p and consequently p is totally ramified in K.

4) If $p \parallel a$, $p^3 \mid b$ and p is totally ramified in K, then $\delta = \alpha/p$ is an integer of K and the minimal polynomial of δ is given by

$$X^3 + a/p^2 \cdot X + b/p^3.$$

But, as $a/p^2 \notin Z$, we have also a contradiction. #

As for the rational prime number 3, we have also the following

Lemma 8. Let k_o be the unique cubic subfield of $Q(\zeta_9)$. Then k_oK is unramified over K if and only if

$$(6.5)\qquad a/3^2 \equiv 2 \pmod 3 \quad (\text{so } 3^2 \,\|\, a), \quad 3^2 \,\|\, b$$

$$\text{or} \quad a \equiv 6 \pmod{3^2}, \quad b \equiv \pm 1 \pmod{3^2}.$$

<u>Proof</u>. First suppose that k_oK is unramified over K; so 3 is totally ramified in K. Then, by (5.13), $3^3 \mid b$ must imply $3^{2+1} = 3^3 \mid a$, which contradicts our assumption (6.2). So we have, by (5.12), $3 \mid a$ and $3^s \,\|\, b$ with $s = 0, 1, 2$.

1) If $s = 1$, then $f(X)$ is of Eisenstein type with respect to 3 but $0 + b = b \not\equiv 0 \pmod{3^2}$. Therefore k_oK is not unramified over K.

2) If $s = 2$, by Example 7, we see that k_oK is unramified over K if and only if

$$3^2 \,\|\, a, \quad a/3^2 \equiv 2 \pmod 3 \quad (3^2 \,\|\, b).$$

3) If $s = 0$, the minimal polynomial of $\alpha + b$ is

$$f(X - b) = X^3 - 3bX^2 + (3b^2 + a)X - (b^3 - b + ab) \in Z[X]$$

with $3 \mid b^3 - b + ab$. Suppose that $3^3 \mid b^3 - b + ab$, then $(\alpha + b)/3$ is an integer of K, whose minimal polynomial is given by

$$X^3 - 3b/3\cdot X^2 + (3b^2 + a)/3^2\cdot X$$
$$- (b^3 - b + ab)/3^3 \in Z[X].$$

But we have $-3b/3 = -b \not\equiv 0 \pmod 3$, which contradicts (5.12). Accordingly we must have $3 \| b^3 - b + ab$ or $3^2 \| b^3 - b + ab$. If $3 \| b^3 - b + ab$, then $f(X - b)$ is of Eisenstein type with respect to 3 and so k_oK is unramified over K if and only if

$$3b^2 + a \equiv -3b - (b^3 - b + ab) \equiv 0 \pmod{3^2}$$
$$\text{i.e.} \quad a \equiv 6 \pmod{3^2},\ b \equiv \pm 1 \pmod{3^2}.$$

If $3^2 \| b^3 - b + ab$, then $3^2 \nmid -3b$ implies that k_oK is not unramified over K (cf. Example 7). Hence if k_oK is unramified over K, we have (6.5).

Conversely suppose that (6.5) is satisfied. Then, as is seen above, the minimal polynomial $h(X)$ of $\alpha^2/3$ for $3^2 \| a$, $3^2 \| b$ or the minimal polynomial $f(X - b)$ of $\alpha + b$ for $a \equiv 6 \pmod{3^2}$, $b \equiv \pm 1 \pmod{3^2}$ is of Eisenstein type with respect to 3. Therefore 3 is totally ramified in K and k_oK is unramified over K. #

Summarizing the above results, we have the following

<u>Theorem 6</u>. Let K be a cubic number field with a primitive

element α and let its minimal polynomial be

$$f(X) = X^3 + aX + b \in Z[X],$$

where there is no prime number p satisfying $p^2 \mid a,\ p^3 \mid b$. Let $p_1, p_2, \dots, p_t$ $(p_i \neq 2, 3)$ be all the rational prime numbers such that

$$p_i \mid a,\ p_i \| b \quad \text{or} \quad p_i^2 \mid a,\ p_i^2 \| b$$
$$\text{and} \quad p_i \equiv 1 \pmod 3.$$

Then we have

$$k_1^* = \prod_{i=1}^{t} k^*(p_i) \qquad \text{(composite)},$$

where $k^*(p_i)$ is the unique cubic subfield of $Q(\zeta_{p_i})$. Also we have

$$k_2^* = \begin{cases} k_0 \ \text{(the unique cubic subfield of } Q(\zeta_9)), \\ \quad \text{if } a/3^2 \equiv 2 \pmod 3,\ 3^2 \| b \\ \qquad \text{or } a \equiv 6 \pmod{3^2},\ b \equiv \pm 1 \pmod{3^2}, \\ Q, \text{ otherwise.} \end{cases}$$

Thus, for the absolute abelian number field $k^* = k_1^* k_2^*$, $K^* =$

k^*K is the genus field of K.

Example 9. 1) Let $K = Q(\eta)$ be a cubic number field, where η is a root of the cubic (irreducible) polynomial

$$X^3 + 6nX - 1 \in Z[X]$$

with $n \in Z$ $(n > 0)$. Then we have shown, in Ishida [17], that if $3 \nmid n$ and $32n^3 + 1$ is squarefree then the class number h_K of K is divisible by 3. As the discriminant of the polynomial is $-4(6n)^3 - 27 = -3^3(32n^3 + 1)$, the prime number 3 is the only prime which is totally ramified in K. Hence, by Theorems 5 and 6, we have

$$g_K = \begin{cases} 3, \text{ if } & 6n \equiv 6 \pmod{3^2} \quad \text{i.e.} \quad n \equiv 1 \pmod 3, \\ 1, \text{ oterwise} & \text{i.e.} \quad n \equiv 2 \pmod 3. \end{cases}$$

This implies that if $n \equiv 2 \pmod 3$ ($32n^3 + 1$ is squarefree) then, in Corollary to Theorem 5, we have a strict inequality

$$d = d^{(3)}C_K > 0.$$

2) A similar example is as follows : Let $K = Q(\sqrt[3]{q_1 q_2 \cdots q_r})$ with $q_i \equiv 2 \pmod 3$. Then, by Theorem 5

and Example 6, we have $g_K = 1$. On the other hand, by Appendix A) to Chapter 2, we have $d = d^{(3)}c_K \geq r - 3$. So if $r > 3$, we have also a strict inequality

$$d = d^{(3)}c_K > 0. \quad \#$$

<u>Example 10</u> (Fröhlich [7]). Let $K = Q(\alpha)$ be a cubic number field with a primitive element α and let D_f be the discriminant of the minimal polynomial $f(X)$ of α. Then we show that the genus number g_K of K is given by

$$g_K = \begin{cases} 3^e, & \text{if } K \text{ is not cyclic over } Q, \\ 3^{e-1}, & \text{if } K \text{ is cyclic over } Q, \end{cases}$$

where e is the number of primes p $(\neq 2)$ such that p is totally ramified in K and $\left(\frac{D_f}{p}\right) = 1$. Here we extend the Legendre-Jacobi symbol for odd prime p by

$$(6.6) \qquad p^s \,\|\, A \implies \begin{cases} \left(\frac{A}{p}\right) = \left(\frac{A/p^s}{p}\right), & \text{if } s \text{ is even}, \\ \left(\frac{A}{p}\right) = 0, & \text{if } s \text{ is odd}. \end{cases}$$

In fact, let

$$g(X) = X^3 + cX + d \in Z[X]$$

be the minimal polynomial of a primitive element β of K without prime number q satisfying $q^2 \mid c$, $q^3 \mid d$. Then, by Lemma 7, if p $(\neq 2, 3)$ is totally ramified in K, we have $p \mid c$, $p \parallel d$ or $p^2 \mid c$, $p^2 \parallel d$. As D_g (the discriminant of $g(X)$) $= -(4c^3 + 27d^2)$, we have

$$\left(\frac{D_g}{p}\right) = \begin{cases} \left(\frac{-27d'^2}{p}\right) = \left(\frac{-3}{p}\right) = \left(\frac{p}{3}\right), \\ \quad \text{if } p \mid c,\ p \parallel d \quad (d' = d/p), \\ \left(\frac{-27d'^2}{p}\right) = \left(\frac{-3}{p}\right) = \left(\frac{p}{3}\right), \\ \quad \text{if } p^2 \mid c,\ p^2 \parallel d \quad (d' = d/p^2). \end{cases}$$

Hence, for a totally ramifying prime p $(\neq 2, 3)$, $p \equiv 1 \pmod 3$ is equivalent to $\left(\frac{D_g}{p}\right) = 1$. On the other hand, if 3 is totally ramified in K, we can take a primitive element γ, whose minimal polynomial

$$h(X) = X^3 + c_1X^2 + c_2X + c_3 \in Z[X]$$

is of Eisenstein type with respect to 3 i.e. we have

$$c_1 = 3d_1,\ c_2 = 3d_2,\ c_3 = 3d_3,\quad 3 \nmid d_3.$$

The discriminant D_h of $h(X)$ is given by

(6.7) $$D_h = c_1{}^2c_2{}^2 + 18c_1c_2c_3 - 4c_2{}^3 - 4c_1{}^3c_3 - 27c_3{}^2$$
$$= 3^4d_1{}^2d_2{}^2 + 2\cdot 3^5d_1d_2d_3 - 4\cdot 3^3d_2{}^3$$
$$- 4\cdot 3^4d_1{}^3d_3 - 3^5d_3{}^2.$$

Then, by comparing the exponents of 3 in each terms, we have

$$3 \nmid d_2 \Longrightarrow 3^3 \| D_h \Longrightarrow \left(\frac{D_k}{3}\right) = 0,$$

$$3 \mid d_2,\ 3 \nmid d_1 + d_3 \text{ and } 3 \mid d_1 \Longrightarrow 3^5 \| D_h$$

$$\Longrightarrow \left(\frac{D_k}{3}\right) = 0,$$

$$3 \mid d_2,\ 3 \nmid d_1 + d_3 \text{ and } 3 \nmid d_1 \quad (\text{so } d_1 \equiv d_3 \pmod 3)$$

$$\Longrightarrow D_h = 3^4(-4d_1{}^3d_3 + 3B) \text{ with } B \in Z$$

$$\Longrightarrow \left(\frac{D_k}{3}\right) = \left(\frac{-d_1d_3}{3}\right) = \left(\frac{-1}{3}\right) = -1,$$

$$3 \mid d_2 \text{ and } 3 \mid d_1 + d_3 \quad (\text{so } 3 \nmid d_1)$$

$$\Longrightarrow D_h = 3^4(-d_1{}^3d_3 + 3B) \text{ with } B \in Z$$

$$\Longrightarrow \left(\frac{D_k}{3}\right) = \left(\frac{-d_1d_3}{3}\right) = \left(\frac{1}{3}\right) = 1.$$

Hence, by Theorem 5, we see that k_oK is unramified over K if and only if $\left(\frac{D_k}{3}\right) = 1$. For a primitive element α of K, the discriminant D_f of the minimal polynomial $f(X)$ of α

is different from D_g and D_h only by a square factor. There fore the assertion is proved. #

Chapter 7. The genus fields of pure number fields

Finally we consider a pure number field $K = Q(\sqrt[n]{a})$ ($a \neq \pm 1$, $\in Z$) and determine the genus field K^* of K, where we assume that n is odd and a has the property

$$(7.1) \qquad p^v \| a \implies (v, n) = 1$$

for any prime divisor p of a.

Suppose, first, that $n = q^s$ is an odd prime power i.e. $K = Q(\sqrt[q^s]{a})$ with an integer $a \neq \pm 1$ such that

$$(7.2) \qquad p^v \| a \implies (v, q) = 1$$

for any prime divisor p of a. Then a rational prime number ramifying in K is either q or a divisor of a. For a rational prime number $p \neq q$ with $p^v \| a$ ($v > 0$, $(v, q) = 1$), there are integers $x, y \in Z$ ($1 \leqq x \leqq q^s - 1$, $y > 0$) such that $vx = 1 + q^s y$ and so we have, as $(q,x) = 1$,

$$K = Q(\sqrt[q^s]{a^x/p^y}) = Q(\sqrt[q^s]{b})$$

with $b \in Z$ and $p \| b$. Hence we see that K is of Eisenstein type with respect to p and p is totally ramified in K i.e.

$e(p) = q^s$ and $p \nmid e(p)$. As for q, $e(q)$ is a divisor of $[K : Q] = q^s$ and so if $e(q) > 1$ then we have $q \mid e(q)$.

Consequently we have, using the notations in Chapter 4,

$$(7.3) \qquad k_1^* = \prod_{p \neq q,\ p \mid a} k^*(p) \qquad \text{(composite)},$$

where $k^*(p)$ is the unique subfield of $Q(\zeta_p)$, of degree $d^*(p) = (e(p), p - 1) = (q^s, p - 1)$.

On the other hand, k_2^* is a subfield of a q-power-th cyclotomic number field and, by considering the degrees, we see that $k_2^* \subset Q(\zeta_{q^{s+1}})$ and $[k_2^* : Q]$ divides q^s. Note that, as q is odd, $Q(\zeta_{q^{s+1}})$ is cyclic over Q. Then we can prove

$$(7.4) \qquad k_2^* = Q.$$

In fact, if $k_2^* \neq Q$, k_2^* must contain the unique subfield k_o, of degree q, of $Q(\zeta_{q^2})$. Then, since $Q(\zeta_{q^2})$ is totally ramified over Q (at q) and $k_o K$ is unramified over K (in particular, at any prime divisor of q in K), there occurs the following situation : for some t $(0 \leq t < s)$, put $M = Q(\alpha)$ with $\alpha = \sqrt[q^t]{a}$. Let $\mathfrak{q}_1, \mathfrak{q}_2, \ldots, \mathfrak{q}_m$ be all the prime divisors of q in M. Then some $\mathfrak{q}_i$ (e.g. $\mathfrak{q}_1$) is totally ramified in $M(\zeta_{q^2})$ and a prime divisor $\mathfrak{q}$ of $\mathfrak{q}_1$ in $M(\sqrt[q]{\alpha})$ is not totally ramified in $L = M(\sqrt[q]{\alpha}, \zeta_{q^2})$. So $\mathfrak{q}_1$ must be totally ramified in $M(\sqrt[q]{\alpha})$: $\mathfrak{q}_1 = \mathfrak{q}^q$. Clearly L is a Galois extension over M.

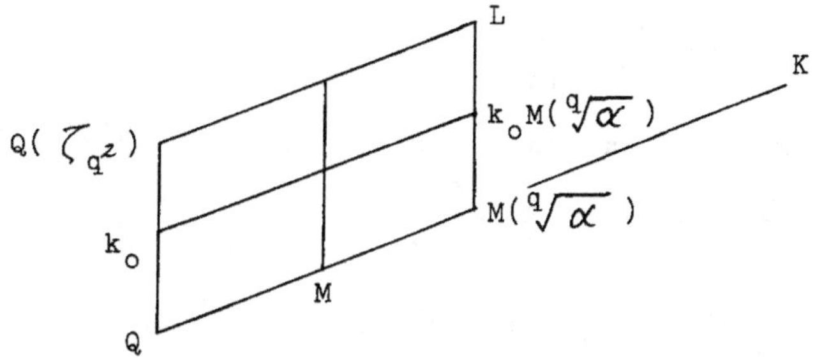

Let σ be a generator of the Galois group of L over $M(\zeta_{q^2})$ and τ a generator of the Galois group of L over $M(\sqrt[q]{\alpha})$; so the Galois group of L over M is generated by σ and τ , where we have

(7.5)
$$\begin{cases} \sigma^q = 1, \quad \tau^{q(q-1)} = 1, \quad \langle\sigma\rangle \cap \langle\tau\rangle = 1 \\ \text{and} \quad \tau\sigma\tau^{-1} = \sigma^r \\ \qquad (r = \text{a primitive root mod } q^2) \\ \sigma : \zeta \longrightarrow \zeta, \ \sqrt[q]{\alpha} \longrightarrow \zeta^q \cdot \sqrt[q]{\alpha}, \\ \tau : \zeta \longrightarrow \zeta^r, \ \sqrt[q]{\alpha} \longrightarrow \sqrt[q]{\alpha} \end{cases}$$
$$(\zeta = \zeta_{q^2}).$$

Then, in the Galois extension L over M, the ramification index of $\mathfrak{q}_1$ is $q(q-1)$ and so the inertia group $\mathfrak{T}$ of a prime divisor $\mathfrak{Q}$ of $\mathfrak{q}$ in L is of order $q(q-1)$. Accordingly $\mathfrak{T}$ contains a subgroup $\mathfrak{X}$ of order q and, as L is totally ramified at prime divisors of $\mathfrak{q}$ in $k_0M(\sqrt[q]{\alpha})$, $\mathfrak{T}$ also contains $\langle \tau^q \rangle$; hence we have $\mathfrak{T} = \mathfrak{X} \langle \tau^q \rangle$. That is, the inertia field T of $\mathfrak{Q}$ in L over M is of relative degree q and is contained in $k_0M(\sqrt[q]{\alpha})$. For an element $\sigma^i \tau^j$ of $\mathfrak{X}$, we have

$$(\sigma^i \tau^j)^q = \sigma^{i(1+r^j+\ldots+r^{(q-1)j})} \tau^{jq} = 1,$$

which implies that $j \equiv 0 \pmod{q-1}$ and so $i(1 + r^j + \ldots + r^{(q-1)j}) \equiv i(1 + 1 + \ldots + 1) = iq \equiv 0 \pmod q$. Hence all the elements of $\mathfrak{X}$ are of the form

$$\sigma^i \tau^{t(q-1)} \qquad (\sigma^i \tau^{t(q-1)} = 1 \iff q \mid i,\ q \mid t).$$

If $q \mid i$ and $q \nmid t$, then $\mathfrak{X}$ contains $\tau^{t(q-1)}$ and so τ^{q-1}. Hence we have $\mathfrak{T} = \langle \tau \rangle$ and $T = M(\sqrt[q]{\alpha})$. If $q \nmid i$ and $q \mid t$, then $\mathfrak{X}$ contains σ^i and therefore σ. Hence $\mathfrak{T} = \langle \sigma \rangle \cdot \langle \tau^q \rangle$ and $T = k_0M$. If $q \nmid i$ and $q \nmid t$, then there are integers $x, y \in Z$ such that $t(q-1)x + qy = 1$. Consequently we have

$$\mathfrak{q} \ni (\sigma^{i}\tau^{t(q-1)})^{x}\tau^{qy}$$
$$= \sigma^{i(1+r^{t(q-1)}+\dots+r^{t(x-1)(q-1)})}\tau^{t(q-1)x+qy}$$
$$= \sigma^{ix}\tau$$

(because $r^{q-1} \equiv 1 \pmod q$). So

$$\mathfrak{q} \ni (\sigma^{ix}\tau)^{q-1} = \sigma^{ix(1+r+\dots+r^{q-2})}\tau^{q-1} = \tau^{q-1}$$

(because $1 + r + \dots + r^{q-2} = (1 - r^{q-1})/(1 - r) \equiv 0 \pmod q$). Hence $\mathfrak{q}$ contains τ and σ^{i} and so σ, which implies that $\mathfrak{q}$ coincides with the whole Galois group of L over M. This is of course a contradiction. Consequently we see that the inertia field T of $\mathfrak{q}$ in L over M is either $M(\sqrt[q]{\alpha})$ or k_oM. On the other hand, by our assumption, the prime ideal $\mathfrak{q}_1$ $(\mathfrak{q} \mid \mathfrak{q}_1)$ of M is totally ramified in $M(\sqrt[q]{\alpha})$ and k_oM. So we have a contradiction. Thus we see that, for a pure number field $K = Q(\sqrt[q^s]{a})$ with (7.2),

$$k_2^* = Q.$$

Accordingly, combining (7.3) with (7.4), we have

$$(7.6) \qquad k^* = \prod_{p \neq q,\ p \mid a} k^*(p) \qquad \text{(composite)},$$

where $k^*(p)$ is the unique subfield, of degree $(q^s, p - 1)$, of $Q(\zeta_p)$. In particular, since K has no absolute abelian subfield, we have the following result of Fröhlich [7] : The genus number g_K of $K = Q(\sqrt[q^s]{a})$ with (7.2) is given by

$$(7.7) \qquad g_K = \prod_{p \mid a} (q^s, p - 1)$$

and the genus group of K is the direct product of cyclic groups of order $(q^s, p - 1)$ for all $p \mid a$.

Now in the case where $n = \prod_{i=1}^{t} q_i^{s_i}$ is an odd integer with primes q_i, we have the following

<u>Theorem 7</u>. Let $K = Q(\sqrt[n]{a})$ be a pure number field, where $n = q_1^{s_1} q_2^{s_2} \ldots q_t^{s_t}$ is odd and a $(\neq \pm 1)$ is an integer having the property such that

$$p^v \parallel a \implies (v, n) = 1$$

for any prime divisor p of a. Then the absolute abelian number field

$$k^* = \prod_{p \neq q_i,\ p \mid a} k_p^* \qquad \text{(composite)}$$

gives the genus field $K^* = k^*K$ of K, where k_p^* is the subfield of $Q(\zeta_p)$, of degree $\prod_{i=1}^{t} (q_i^{s_i}, p - 1) = (n, p - 1)$.

In particular, the genus number g_K of K is given by

$$g_K = \prod_{p \neq q_i,\ p | a} (n, p - 1)$$

and the genus group of K is the direct product of cyclic groups of order $(n, p - 1)$ for all p such that $p \neq q_i$, $p | a$.

Proof. Obviously $Q(\sqrt[n]{a}) \supset \prod_{i=1}^{t} Q(\sqrt[q_i^{s_i}]{a})$ (composite). On the other hand, there are integers $u_i \in Z$ such that $\sum_{i=1}^{t} n/q_i^{s_i} \cdot u_i = 1$. So $\prod_{i=1}^{t} (\sqrt[q_i^{s_i}]{a})^{u_i} = \sqrt[n]{a}$, which implies

$$Q(\sqrt[n]{a}) = \prod_{i=1}^{t} Q(\sqrt[q_i^{s_i}]{a}) \quad \text{(composite)}.$$

Since the degrees $[Q(\sqrt[q_i^{s_i}]{a}) : Q] = q_i^{s_i}$ are coprime to each other, the assertion follows from Proposition 2 and the preceding results. #

Here we give another simple proof of (7.4) under some assump tion : Let $K = Q(\sqrt[q^s]{a})$, where q is an odd prime and $a \neq \pm 1$ satisfies (7.2). (If q is unramified in K, then we have of course $k_2^* = Q$.) Now we prove that if q is totally ramified in the subfield $Q(\sqrt[q]{a})$ of degree q, we have also $k_2^* = Q$. In fact, as remarked above, we note that $k_2^* \neq Q$ implies $k_2^* \supset k_o$, where k_o is the unique subfield, of degree q, of $Q(\zeta_{q^2})$. Also note that there is no ramification of infinite prime divisors, in k_oK, of K. Clearly k_o corresponds to the ideal group

$$H_{q^2} = \left\{ (a) \in A_q \;\middle|\; a^{q-1} \equiv 1 \pmod{^{\times}\, q^2} \right\}$$

(cf. Chapter 5). Accordingly, if we show that there is an integer γ, prime to q, of K such that

$$N_K \gamma^{q-1} \not\equiv 1 \pmod{q^2},$$

then we have $k_2^* \not\supset k_o$ and consequently $k_2^* = Q$.

1) If $q \mid a$, similarly as in the beginning of this chapter, we have

$$K = Q(\sqrt[q^s]{b}), \quad q \,\|\, b$$

for some $b \in Z$. Put $\gamma = 1 + \sqrt[q^s]{b}$. Then, from $(\gamma - 1)^{q^s} = b$, it follows

$$N_K\gamma = 1 + b \quad \text{and so}$$

$$N_K\gamma^{q-1} \equiv 1 + (q - 1)b \equiv 1 - b \not\equiv 1 \pmod{q^2}.$$

2) If $q \nmid a$, by our assumption, we have $q \parallel a^q - a$ i.e. $q^2 \nmid a^{q-1} - 1$ (cf. Chapter 5). Put $\gamma = \sqrt[q^s]{a}$. Then, from $\gamma^{q^s} = a$, it follows

$$N_K\gamma = a \quad \text{and so}$$

$$N_K\gamma^{q-1} = a^{q-1} \not\equiv 1 \pmod{q^2}.$$

We can apply a similar reasoning to the case $q = 2$ and $2 \parallel a$ as follows : Let $K = Q(\sqrt[2^s]{a})$ with (7.2) (for $q = 2$). Then, as in the beginning of this chapter, we have

$$k_1^* = \prod_{p \neq 2,\ p | a} k^*(p) \qquad \text{(composite)},$$

where $k^*(p)$ is the unique subfield, of degree $(2^s, p - 1)$, of $Q(\zeta_p)$. Now we prove

$$k_2^* = \begin{cases} Q(\sqrt{2}), & \text{if } a \equiv 2 \pmod 8, \\ Q(\sqrt{-2}), & \text{if } a \equiv -2 \pmod 8. \end{cases}$$

In fact, for the case $a \equiv 2 \pmod 8$, $Q(\sqrt{2})Q(\sqrt{a})$ is unramified over $Q(\sqrt{a})$ (cf. Chapter 4) and so we have $k_2^* \supset Q(\sqrt{2})$. Consequently $k_2^* \neq Q(\sqrt{2})$ $(a \equiv 2 \pmod 8)$ implies $k_2^* \supset Q(\zeta_8)$. Clearly $Q(\zeta_8)$ corresponds to the ideal group

$$S_{\tilde{8}} = \left\{ (a) \in A_2 \;\middle|\; a \equiv 1 \pmod{^\times \tilde{8}} \right\}$$

with defining modulus $\tilde{8} = 8p_\infty$. Put $\gamma = (a + 1) + \sqrt[2^s]{a}$. Then γ is a totally positive integer of K, prime to 2, and we have

$$N_K \gamma = (a + 1)^{2^s} - a \equiv 7 \not\equiv 1 \pmod{^\times \tilde{8}}.$$

This implies that $Q(\zeta_8)K$ is not unramified over K and so $k_2^* \not\supset Q(\zeta_8)$ i.e. $k_2^* = Q(\sqrt{2})$.

Notations

Z is the ring of rational integers and Q is the field of rational numbers.

Let $a \in Z$ and p a rational prime number. If p is a divisor of a, we write $p \mid a$ ($p \nmid a$ in negative case). Moreover if p^e is the exact power of p dividing a, then we write $p^e \parallel a$.

Let $a, b \in Z$. Then (a, b) denotes the greatest common divisor of a and b.

Let p_∞ be the (unique) infinite prime divisor of Q. For an integral divisor $\tilde{m}$ of Q (i.e. $\tilde{m} = m$ or mp_∞ with some $m \in Z$ $(m > 0)$), $A_{\tilde{m}}$ denotes the group of all (principal) ideals, prime to m, in Q and $S_{\tilde{m}}$ the 'Strahl mod $\tilde{m}$' in Q i.e. the subgroup of $A_{\tilde{m}}$ consisting of all ideals (a) with $a \equiv 1 \pmod{^{\times} \tilde{m}}$ (multiplicative congruence). A subgroup $H_{\tilde{m}}$ of $A_{\tilde{m}}$, containing $S_{\tilde{m}}$, is called an ideal group with defining modulus $\tilde{m}$.

For a positive rational integer $m \in Z$, ζ_m denotes a primitive m-th root of unity. Then $Q(\zeta_m)$ is the m-th cyclotomic number field and $Q(\zeta_m)_o$ is its maximal real subfield.

Let K be an algebraic number field of finite degree. Then D_K denotes the discriminant of K, O_K the ring of integers in K. The norm mapping defined for elements and ideals of K is denoted by N_K. Moreover C_K is the ideal class group and h_K is the class number of K. (When we consider the ideal class group or the class number in 'narrow' sense, we denote them by C_K^+ or h_K^+ respectively.) For a prime number q, $d^{(q)}C_K$ is the q-rank of C_K. i.e. the number of invariants of C_K which is a power of q.

Let p be a rational prime number and let

$$(p) = \mathfrak{p}_1^{e_1} \mathfrak{p}_2^{e_2} \cdots \mathfrak{p}_m^{e_m}$$

be the prime ideal decomposition of p in K. Then $e(p)$ is the greatest common divisor of $e_1, e_2, \ldots, e_m$ and $d^*(p) = (e(p), p-1)$. Moreover $k^*(p)$ is the unique subfield, of degree $d^*(p)$, of $Q(\zeta_p)$.

For an algebraic number field K, K^* is the genus field of K, $g_K = [K^* : K]$ the genus number of K and k^* the maximal absolute abelian subfield of K^* (as for the definitions, see Chapters 1 and 4).

Finally, for several algebraic number fields $K_1, K_2, \ldots,$ K_t,

$$\prod_{i=1}^{t} K_i = K_1K_2 \ldots K_t$$

denotes the composite field of them i.e. the smallest number field containing all of them.

References

[1] Z.I.Borevich and I.R.Shafarevich, 'Number theory' (English translation), Academic Press, New York, 1966.

[2] P.Barrucand and H.Cohn, A rational genus, class number divisibility and unit theorem for pure cubic fields, J. Number Theory 2 (1970), 7-21.

[3] H.Cohn, 'A second course in number theory', Wiley, New York, 1962.

[4] I.Connell and D.Sussman, The p-dimension of class groups of number fields, J. London Math. Society (second series) 2 (1970), 525-529.

[5] G.Frey and W.D.Geyer, Über die Fundamentalgruppe von Körpern mit Divisorentheorie, J. Reine Angew. Math. 254 (1972), 110-122.

[6] A.Fröhlich, The genus field and genus group in finite number fields, Mathematika 6 (1959), 40-46.

[7] A.Fröhlich, The genus field and genus group in finite number field, II, Mathematika 6 (1959), 142-146.

[8] Y.Furuta, The genus field and genus number in algebraic number field, Nagoya Math. J. 29 (1967), 281-285.

[9] Y.Furuta, Über das Geschlecht und die Klassenzahl eines relativ-Galoissches Zahlkörpers von Primzahlpotenzgrade, Nagoya Math. J. 37 (1970), 197-200.

[10] H.Hasse, 'Vorlesungen über Klassenkörpertheorie', Physica Verlag, Würzburg, 1967.

[11] H.Hasse, 'Zahlentheorie', Akademie-Verlag, Berlin, 1969.

[12] H.Hasse, Zur Geschlechtertheorie in quadratischen Zahlkörpern, J. Math. Soc. Japan 3 (1951), 45-51.

[13] H.Hasse, A supplement to Leopoldt's theory of genera in abelian number fields, J. Number Theory 1 (1969), 4-7.

[14] M.Ishida, A note on class numbers of algebraic number fields, J. Number Theory 1 (1969), 65-69.

[15] M.Ishida, Class numbers of algebraic number fields of Eisenstein type, J. Number Theory 2 (1970), 404-413.

[16] M.Ishida, Class numbers of algebraic number fields of Eisenstein type, II, J. Number Theory 6 (1974), 99-104.

[17] M.Ishida, Fundamental units of certain algebraic number fields, Abh. Math. Sem. Univ. Hamburg 39 (1973), 245-250.

[18] M.Ishida, Some unramified abelian extensions of algebraic number fields, J. Reine Angew. Math. 268/269 (1974), 165-173.

[19] M.Ishida, On the genus field of an algebraic number field of odd prime degree, J. Math. Soc. Japan 27 (1975), 289-293.

[20] M.Ishida, On the genus field of an algebraic number field of odd prime degree, II (in Japanese), preprint.

[21] M.Ishida, 'Algebraic number theory' (in Japanese), Morikita, Tokyo, 1974.

[22] S.Iyanaga and T.Tamagawa, Sur la théorie du corps de classes sur le corps des nombres rationnels, J. Math. Soc. Japan 3 (1951), 220-227.

[23] H.W.Leopoldt, Zur Geschlechtertheorie in abelschen Zahlkörpern, Math. Nachr. 9 (1953), 350-362.

[24] M.L.Madan, Class groups of global fields, J. Reine Angew. Math. 252 (1972), 171-177.

[25] P.Roquette and H.Zassenhaus, A class rank estimate for algebraic number fields, J. London Math. Soc. 44 (1969), 31-38.

[26] T.Takagi, 'Algebraic number theory' (in Japanese), Iwanami, Tokyo, 1948.

Appendix. An algorithm for constructing the genus field in the case of odd prime degree

In this appendix, we extend the result of Chapter 6 to the case of arbitrary odd prime degree. Namely, we show that we can determine the genus field of an algebraic number field of odd prime degree explicitly from the coefficients of the minimal polynomial of a primitive element by the method obtained in Chapter 5. As for the details of the proofs, see the forthcoming paper in Jour. Fac. Sci. Univ. of Tokyo (vol. 24-1).

Let K be an algebraic number field of odd prime degree q. We keep the notations in Chapter 5 ; in particular, let

$k^*(p)$ = the unique subfield, of degree q, of $Q(\zeta_p)$, for $p \equiv 1 \pmod q$,

$k_1^* = \prod_{i=1}^{t} k^*(p_i)$, where $p_1, p_2, \dots, p_t$ are all the totally ramifying rational prime numbers (in K) such that $p_i \equiv 1 \pmod q$,

k_o = the unique subfield, of degree q, of $Q(\zeta_{q^2})$.

Then, by Theorem 5, the maximal absolute abelian subfield k^* of the genus field $K^* = k^*K$ of K is given by

(A1) $$k^* = k_1^* \, k_2^*,$$

where $k_2^* = Q$ or k_o. The case $k_2^* = k_o$ occurs if and only if the following two conditions are satisfied :

1) q is totally ramified in K. (Accordingly, there is a primitive element π of K such that the minimal polynomial $f(X) = X^q + a_1X^{q-1} + \ldots + a_q \in Z[X]$ of π is of Eisenstein type with respect to q. Then)

2) $$a_1 + a_q \equiv a_2 \equiv \ldots \equiv a_{q-1} \equiv 0 \pmod{q^2}.$$

Now we explain our method to determine the subfield k^*. Suppose that there is given the minimal polynomial

$$X^q + d_1X^{q-1} + \ldots + d_q \in Z[X]$$

of a primitive element δ of K. Then $\beta = q\delta + d_1$ is also a primitive element of K and the minimal polynomial of β is of the form

(A2) $$g(X) = X^q + b_2X^{q-2} + \ldots + b_q \in Z[X].$$

Of course, the coefficients $b_2, \ldots, b_q$ of $g(X)$ are easily calculated from $d_1, d_2, \ldots, d_q$. Moreover, without loss of generality, we may assume that

(A3) there is no prime p such that $p^i \mid b_i$ $(i=2,\dots,q)$.

In fact, if there is such a rational prime number p, we can replace β by β/p.

Therefore we may start with the minimal polynomial $g(X)$ (of a primitive element of K), which is of the form (A2) and has the property (A3).

In this case, we see that

(A4) $p \neq q$ is totally ramified in K

$$\Longrightarrow \quad b_2 \equiv \dots \equiv b_q \equiv 0 \pmod{p} \quad \text{and} \quad p^s \parallel b_q \quad \text{with} \quad 1 \leqq s \leqq q-1,$$

(A5) q is totally ramified in K

$$\Longrightarrow \quad b_2 \equiv \dots \equiv b_{q-1} \equiv 0 \pmod{q} \quad \text{and} \quad q^s \parallel b_q \quad \text{with} \quad 0 \leqq s \leqq q-1.$$

Then we can know all the primes, totally ramifying in K, and can decide whether $k_2{}^* = k_o$ or not in the following way (cf. the latter half of Chapter 5).

First, suppose that a rational prime number p $(\neq q)$ satisfies the conditions (cf. (A4))

(*) $$b_2 \equiv \dots \equiv b_q \equiv 0 \pmod{p} \quad \text{and} \quad p^s \parallel b_q \quad \text{with} \quad 1 \leqq s \leqq q-1$$

Then there exist $m, n \in Z$ such that $ms = 1 + nq$ $(1 \leqq m \leqq q - 1, n \geqq 0)$ and $\gamma = \beta^m/p^n$ is a primitive element of K. Let $h(X) = X^q + c_1X^{q-1} + \dots + c_q \in Q[X]$ be the minimal polynomial of γ, where $c_q \in Z$ and $p \parallel c_q$.

<u>Lemma A1</u>. Under the conditions (*), p is totally ramified in K

$$\Longleftrightarrow \quad c_1 \equiv c_2 \equiv \dots \equiv c_{q-1} \equiv 0 \pmod{p}.$$

Next, suppose that q satisfies the conditions (cf. (A5))

(**) $\qquad b_2 \equiv \dots \equiv b_{q-1} \equiv 0 \pmod{q}$ and $q^s \,|\, b_q$ with $0 \leqq s \leqq q-1$.

We consider two cases $s \geqq 1$ and $s = 0$ separately. In case $s \geqq 1$, similarly as above, for $m, n \in Z$ with $ms = 1 + nq$ $(1 \leqq m \leqq q - 1, n \geqq 0)$, let $t(X) = X^q + r_1X^{q-1} + \dots + r_q \in Q[X]$ be the minimal polynomial of $\rho = \beta^m/q^n$, where $r_q \in Z$ and $q \parallel r_q$.

<u>Lemma A2</u>. Under the conditions (**) with $s \geqq 1$, we have $k_2^* = k_o$ (i.e. k_oK is unramified over K)

$$\Longleftrightarrow \quad r_1 + r_q \equiv r_2 \equiv \dots \equiv r_{q-1} \equiv 0 \pmod{q^2}.$$

Since $q \parallel r_q$, the first congruence $r_1 + r_q \equiv 0 \pmod{q^2}$ holds if and only if

$$q \parallel r_1 \quad \text{and} \quad \frac{r_1}{q} + \frac{r_q}{q} \equiv 0 \pmod{q}.$$

In case $s = 0$ i.e. $q \nmid b_q$, we replace β by $\beta' = \beta + b_q$. Then the minimal polynomial of β' is

$$g'(X) = g(X - b_q) = X^q + b_1'X^{q-1} + \ldots + b_q' \in Z[X].$$

If q is totally ramified in K, then we have $q^{s'} \parallel b_q'$ with $1 \leq s' \leq q - 1$. There are also $m', n' \in Z$ such that $m's' = 1 + n'q$ $(1 \leq m' \leq q - 1,\ n' \geq 0)$. Let $t'(X) = X^q + r_1'X^{q-1} + \ldots + r_q' \in Q[X]$ be the minimal polynomial of $\rho' = \beta'^{m'}/q^{n'}$, where $r_q' \in Z$ and $q \parallel r_q'$.

Lemma A3. Under the conditions (**) with $s = 0$, we have $k_2^* = k_o$ (i.e. k_oK is unramified over K)

$$\Longleftrightarrow \quad q^{s'} \parallel b_q' \text{ with } 1 \leq s' \leq q - 1 \text{ and}$$
$$r_1' + r_q' \equiv r_2' \equiv \ldots \equiv r_{q-1}' \equiv 0 \pmod{q^2}.$$

Note that the coefficients c_i, r_i and r_i' of $h(X)$, $t(X)$ and $t'(X)$ can be calculated from b_j in (A2) in elementary way.

Therefore, by (A1) and by Lemmas A1, A2, A3, we can determine k^* and so $K^* = k^*K$ from the coefficients $b_2, \ldots, b_q$ of $g(X)$. Thus we obtain an algorithm for constructing the genus field K^*

of K.

Here we restate our algorithm in more explicit form.

First we introduce a notation. Let q be an odd prime number and let α be an algebraic integer, of degree q, with the minimal polynomial

$$(A6) \qquad f(X) = X^q + a_1X^{q-1} + \dots + a_q \in Z[X].$$

Then, for $m \in Z$ $(1 \leqq m \leqq q - 1)$, α^m is also an algebraic integer, of degree q. Let

$$(A7) \qquad f_m(X) = X^q + a_1^{(m)}X^{q-1} + \dots + a_q^{(m)} \in Z[X]$$

be the minimal polynomial of α^m. By the fundamental theorem on symmetric polynomials, the coefficients $a_1^{(m)}$, $a_2^{(m)}$, ... , $a_q^{(m)}$ of $f_m(X)$ can be calculated from a_1, a_2, ... , a_q in elementary (but somewhat complicated) way.

<u>The algorithm</u>

Let K be an algebraic number field, of odd prime degree q, and let

$$(A8) \qquad g(X) = X^q + b_2X^{q-2} + \dots + b_q \in Z[X]$$

be the minimal polynomial of a primitive element β of K with the property

(A9) there is no prime p such that $p^i \mid b_i$ $(i=2,\ldots,q)$.

Then the maximal absolute abelian subfield k^* of the genus field K^* of K is given by

(A10) $$k^* = \prod_{p \equiv 1 \pmod q} k^*(p)^{\delta_p} \, k_o^{\delta_q} \quad \text{(composite)}$$

with δ_p, $\delta_q = 0$ or 1,

where the first product (composite) ranges over all the rational prime numbers $p \equiv 1 \pmod q$. Here, for an algebraic number field k, we denote

$$k^0 = Q \quad \text{and} \quad k^1 = k.$$

The 'exponents' δ_p and δ_q are determined as follows.

(I) $p \neq q$ $(p \equiv 1 \pmod q)$

We have $\delta_p = 1$ if and only if

(1a) $b_2 \equiv \ldots \equiv b_q \equiv 0 \pmod p$,

(1b) $p^s \| b_q$ with $1 \leqq s \leqq q-1$,

(1c) for $m, n \in Z$ with $ms = 1 + nq$ $(1 \leqq m \leqq q-1,\ n \geqq 0)$,

$$b_1^{(m)} \equiv 0 \pmod{p^{n+1}}, \ \ldots\ , b_{q-1}^{(m)} \equiv 0 \pmod{p^{(q-1)n+1}}.$$

(II) $p = q$

We have $\delta_q = 1$ if and only if

(2a) $b_2 \equiv \ldots \equiv b_{q-1} \equiv 0 \pmod{q}$,

(2b) $q^s \| b_q$ with $0 \leqq s \leqq q-1$,

($2c_1$) when $s \geqq 1$, for $m, n \in Z$ with $ms = 1 + nq$ $(1 \leqq m \leqq q-1,\ n \geqq 0)$,

$$\begin{cases} q^{n+1} \| b_1^{(m)}, \quad \dfrac{b_1^{(m)}}{q^{n+1}} + \dfrac{b_q^{(m)}}{q^{qn+1}} \equiv 0 \pmod{q}, \\ b_2^{(m)} \equiv 0 \pmod{q^{2n+2}}, \ \ldots\ , b_{q-1}^{(m)} \equiv 0 \pmod{q^{(q-1)n+2}}. \end{cases}$$

($2c_2$) when $s = 0$, for $g'(X) = g(X - b_q) = X^q + b_1'X^{q-1} + \ldots + b_q' \in Z[X]$,

$q^{s'} \| b_q' = g(-b_q)$ with $1 \leqq s' \leqq q-1$

and, for $m', n' \in Z$ with $m's = 1 + n'q$ $(1 \leqq m' \leqq q-1,\ n' \geqq 0)$,

$$\begin{cases} q^{n'+1} \,\|\, b_1{}'^{(m')}, \quad \dfrac{b_1{}'^{(m')}}{q^{n'+1}} + \dfrac{b_q{}'^{(m')}}{q^{qn'+1}} \equiv 0 \pmod q \\ b_2{}'^{(m')} \equiv 0 \pmod{q^{2n'+2}}, \ \ldots, \\ \qquad b_{q-1}{}'^{(m')} \equiv 0 \pmod{q^{(q-1)n'+2}}. \end{cases}$$

In the following, we apply our algorithm to polynomials of special type.

A) Trinomial case

We consider an irreducible trinomial

(A11) $$g(X) = X^q + aX + b \in Z[X]$$

of odd prime degree q, having the property

(A12) there is no prime p such that $p^{q-1} \mid a$ and $p^q \mid b$.

Let β be a root of $g(X)$ and let $K = Q(\beta)$ be an algebraic number field generated by β over Q. Of course, the degree of K is equal to q. Then we can determine the 'exponents' δ_p and δ_q in

$$k^* = \prod_{p \equiv 1 \pmod q} k^*(p)^{\delta_p} \cdot k_o^{\delta_q}.$$

That is, we have

$$\delta_p = 1 \quad (p \equiv 1 \pmod q)$$

$$\Longleftrightarrow \quad p^s \| b,\ p^s \mid a \quad (s = 1, 2, \dots, q-1),$$

$$\delta_q = 1$$

$$\Longleftrightarrow \begin{cases} q^{q-1} \| b,\ q^{q-1} \| a \ \text{ and } \ \dfrac{a}{q^{q-1}} \equiv -1 \pmod q, \\ \text{or} \\ q = 3 \ \text{ and } \ a \equiv 6,\ b \equiv \pm 1 \pmod{3^2}. \end{cases}$$

Thus the genus field $K^* = k^*K$ of K (defined by an irreducible trinomial (A11)) is completely and explicitly determined. In particular, the genus field of a cubic number field is determined (cf. Theorem 6).

Here we give several numerical examples in cubic case. Let $K = Q(\beta)$ with $\beta^3 + a\beta + b = 0$ be a cubic number field.

a	b	D_K	h_K	δ_p $(p \neq 3)$	δ_3	k*	g_K
-5	5	$-5^2 \cdot 7$	1	0	0	Q	1
4	6	$-2^2\ 307$	3	0	0	Q	1
0	7	$-3^3 \cdot 7^2$	3	$\delta_7=1$, $\delta_p=0$ $(p \neq 7)$	0	$k^*(7)$	3
6	8	$-2^3\ 3^4$	3	0	1	k_0	3
-9	9	3^4	1	0	1	k_0	1

In the last example, $K = k_o$ is cyclic over Q. (The values of D_K and h_K are taken from the paper of Reid in Amer. Jour. of Math. 23 (1901).)

B) Quintic case

Let K be an algebraic number field of degree 5 and let β be a primitive element of K, where the minimal polynomial of β is of the form

(A13) $$g(X) = X^5 + b_2X^3 + \ldots + b_5 \in Z[X]$$

and has the property

(a14) there is no prime p such that $p^i \mid b_i \quad (i=2,\ldots,5)$

Then we can determine the 'exponents' δ_p and δ_5 in

$$k^* = \prod_{p \equiv 1 \pmod 5} k^*(p)^{\delta_p} \cdot k_o^{\delta_5}.$$

That is, we have the following tables (cf. Chapter 5).

$\delta_p = 1 \quad (p \equiv 1 \pmod 5)$

$\Longleftrightarrow$

the highest exponent of p in			
b_5	b_2	b_3	b_4
1	$\geqq 1$	$\geqq 1$	$\geqq 1$
2	$\geqq 1$	$\geqq 2$	$\geqq 2$
3	$\geqq 2$	$\geqq 2$	$\geqq 3$
4	$\geqq 2$	$\geqq 3$	$\geqq 4$

$\delta_5 = 1$

$\Longleftrightarrow$

the highest exponent of 5 in				
b_5	b_2	b_3	b_4	with the relation
2	$\geqq 2$	2	$\geqq 3$	$\frac{b_3}{5^2} \cdot \frac{b_5}{5^2} \equiv 3 \pmod 5$
3	2	$\geqq 3$	$\geqq 4$	$\frac{b_2}{5^2} \cdot (\frac{b_5}{5^3})^2 \equiv 3 \pmod 5$
4	$\geqq 3$	$\geqq 4$	4	$\frac{b_4}{5^4} \equiv -1 \pmod 5$
0	the highest exponent of 5 in			

the highest exponent of 5 in					
b_5'	b_1'	b_2'	b_3'	b_4'	with the relation
1	1	$\geqq 2$	$\geqq 2$	$\geqq 2$	$\frac{b_1'}{5} \equiv -\frac{b_5'}{5} \pmod 5$

(Here $g'(X) = g(X - b_5) = X^5 + b_1'X^4 + \ldots + b_5' \in Z[X]$.)

As a remark, the conditions of the last row in the table for δ_5 are satisfied if and only if

$b_5 \pmod{5^2}$	$b_2 \pmod{5^2}$	$b_3 \pmod{5^2}$	$b_4 \pmod{5^2}$
± 1	-10	± 5	10
± 7	10	∓ 10	10

Thus the genus field $K^* = k^*K$ of an algebraic number field K, of degree 5, is completely and explicitly determined.

Subject Index

Vol. 399: Functional Analysis and its Applications. Proceedings 1973. Edited by H. G. Garnir, K. R. Unni and J. H. Williamson. II, 584 pages. 1974.

Vol. 400: A Crash Course on Kleinian Groups. Proceedings 1974. Edited by L. Bers and I. Kra. VII, 130 pages. 1974.

Vol. 401: M. F. Atiyah, Elliptic Operators and Compact Groups. V, 93 pages. 1974.

Vol. 402: M. Waldschmidt, Nombres Transcendants. VIII, 277 pages. 1974.

Vol. 403: Combinatorial Mathematics. Proceedings 1972. Edited by D. A. Holton. VIII, 148 pages. 1974.

Vol. 404: Théorie du Potentiel et Analyse Harmonique. Edité par J. Faraut. V, 245 pages. 1974.

Vol. 405: K. J. Devlin and H. Johnsbråten, The Souslin Problem. VIII, 132 pages. 1974.

Vol. 406: Graphs and Combinatorics. Proceedings 1973. Edited by R. A. Bari and F. Harary. VIII, 355 pages. 1974.

Vol. 407: P. Berthelot, Cohomologie Cristalline des Schémas de Caracteristique p > o. II, 604 pages. 1974.

Vol. 408: J. Wermer, Potential Theory. VIII, 146 pages. 1974.

Vol. 409: Fonctions de Plusieurs Variables Complexes, Séminaire François Norguet 1970–1973. XIII, 612 pages. 1974.

Vol. 410: Séminaire Pierre Lelong (Analyse) Année 1972–1973. VI, 181 pages. 1974.

Vol. 411: Hypergraph Seminar. Ohio State University, 1972. Edited by C. Berge and D. Ray-Chaudhuri. IX, 287 pages. 1974.

Vol. 412: Classification of Algebraic Varieties and Compact Complex Manifolds. Proceedings 1974. Edited by H. Popp. V, 333 pages. 1974.

Vol. 413: M. Bruneau, Variation Totale d'une Fonction. XIV, 332 pages. 1974.

Vol. 414: T. Kambayashi, M. Miyanishi and M. Takeuchi, Unipotent Algebraic Groups. VI, 165 pages. 1974.

Vol. 415: Ordinary and Partial Differential Equations. Proceedings 1974. XVII, 447 pages. 1974.

Vol. 416: M. E. Taylor, Pseudo Differential Operators. IV, 155 pages. 1974.

Vol. 417: H. H. Keller, Differential Calculus in Locally Convex Spaces. XVI, 131 pages. 1974.

Vol. 418: Localization in Group Theory and Homotopy Theory and Related Topics. Battelle Seattle 1974 Seminar. Edited by P. J. Hilton. VI, 172 pages 1974.

Vol. 419: Topics in Analysis. Proceedings 1970. Edited by O. E. Lehto, I. S. Louhivaara, and R. H. Nevanlinna. XIII, 392 pages. 1974.

Vol. 420: Category Seminar. Proceedings 1972/73. Edited by G. M. Kelly. VI, 375 pages. 1974.

Vol. 421: V. Poénaru, Groupes Discrets. VI, 216 pages. 1974.

Vol. 422: J.-M. Lemaire, Algèbres Connexes et Homologie des Espaces de Lacets. XIV, 133 pages. 1974.

Vol. 423: S. S. Abhyankar and A. M. Sathaye, Geometric Theory of Algebraic Space Curves. XIV, 302 pages. 1974.

Vol. 424: L. Weiss and J. Wolfowitz, Maximum Probability Estimators and Related Topics. V, 106 pages. 1974.

Vol. 425: P. R. Chernoff and J. E. Marsden, Properties of Infinite Dimensional Hamiltonian Systems. IV, 160 pages. 1974.

Vol. 426: M. L. Silverstein, Symmetric Markov Processes. X, 287 pages. 1974.

Vol. 427: H. Omori, Infinite Dimensional Lie Transformation Groups. XII, 149 pages. 1974.

Vol. 428: Algebraic and Geometrical Methods in Topology, Proceedings 1973. Edited by L. F. McAuley. XI, 280 pages. 1974.

Vol. 429: L. Cohn, Analytic Theory of the Harish-Chandra C-Function. III, 154 pages. 1974.

Vol. 430: Constructive and Computational Methods for Differential and Integral Equations. Proceedings 1974. Edited by D. L. Colton and R. P. Gilbert. VII, 476 pages. 1974.

Vol. 431: Séminaire Bourbaki – vol. 1973/74. Exposés 436–452. IV, 347 pages. 1975.

Vol. 432: R. P. Pflug, Holomorphiegebiete, pseudokonvexe Gebiete und das Levi-Problem. VI, 210 Seiten. 1975.

Vol. 433: W. G. Faris, Self-Adjoint Operators. VII, 115 pages. 1975.

Vol. 434: P. Brenner, V. Thomée, and L. B. Wahlbin, Besov Spaces and Applications to Difference Methods for Initial Value Problems. II, 154 pages. 1975.

Vol. 435: C. F. Dunkl and D. E. Ramirez, Representations of Commutative Semitopological Semigroups. VI, 181 pages. 1975.

Vol. 436: L. Auslander and R. Tolimieri, Abelian Harmonic Analysis, Theta Functions and Function Algebras on a Nilmanifold. V, 99 pages. 1975.

Vol. 437: D. W. Masser, Elliptic Functions and Transcendence. XIV, 143 pages. 1975.

Vol. 438: Geometric Topology. Proceedings 1974. Edited by L. C. Glaser and T. B. Rushing. X, 459 pages. 1975.

Vol. 439: K. Ueno, Classification Theory of Algebraic Varieties and Compact Complex Spaces. XIX, 278 pages. 1975

Vol. 440: R. K. Getoor, Markov Processes: Ray Processes and Right Processes. V, 118 pages. 1975.

Vol. 441: N. Jacobson, PI-Algebras. An Introduction. V, 115 pages. 1975.

Vol. 442: C. H. Wilcox, Scattering Theory for the d'Alembert Equation in Exterior Domains. III, 184 pages. 1975.

Vol. 443: M. Lazard, Commutative Formal Groups. II, 236 pages. 1975.

Vol. 444: F. van Oystaeyen, Prime Spectra in Non-Commutative Algebra. V, 128 pages. 1975.

Vol. 445: Model Theory and Topoi. Edited by F. W. Lawvere, C. Maurer, and G. C. Wraith. III, 354 pages. 1975.

Vol. 446: Partial Differential Equations and Related Topics. Proceedings 1974. Edited by J. A. Goldstein. IV, 389 pages. 1975.

Vol. 447: S. Toledo, Tableau Systems for First Order Number Theory and Certain Higher Order Theories. III, 339 pages. 1975.

Vol. 448: Spectral Theory and Differential Equations. Proceedings 1974. Edited by W. N. Everitt. XII, 321 pages. 1975.

Vol. 449: Hyperfunctions and Theoretical Physics. Proceedings 1973. Edited by F. Pham. IV, 218 pages. 1975.

Vol. 450: Algebra and Logic. Proceedings 1974. Edited by J. N. Crossley. VIII, 307 pages. 1975.

Vol. 451: Probabilistic Methods in Differential Equations. Proceedings 1974. Edited by M. A. Pinsky. VII, 162 pages. 1975.

Vol. 452: Combinatorial Mathematics III. Proceedings 1974. Edited by Anne Penfold Street and W. D. Wallis. IX, 233 pages. 1975.

Vol. 453: Logic Colloquium. Symposium on Logic Held at Boston, 1972–73. Edited by R. Parikh. IV, 251 pages. 1975.

Vol. 454: J. Hirschfeld and W. H. Wheeler, Forcing, Arithmetic, Division Rings. VII, 266 pages. 1975.

Vol. 455: H. Kraft, Kommutative algebraische Gruppen und Ringe. III, 163 Seiten. 1975.

Vol. 456: R. M. Fossum, P. A. Griffith, and I. Reiten, Trivial Extensions of Abelian Categories. Homological Algebra of Trivial Extensions of Abelian Categories with Applications to Ring Theory. XI, 122 pages. 1975.

Vol. 457: Fractional Calculus and Its Applications. Proceedings 1974. Edited by B. Ross. VI, 381 pages. 1975.

Vol. 458: P. Walters, Ergodic Theory – Introductory Lectures. VI, 198 pages. 1975.

Vol. 459: Fourier Integral Operators and Partial Differential Equations. Proceedings 1974. Edited by J. Chazarain. VI, 372 pages. 1975.

Vol. 460: O. Loos, Jordan Pairs. XVI, 218 pages. 1975.

Vol. 461: Computational Mechanics. Proceedings 1974. Edited by J. T. Oden. VII, 328 pages. 1975.

Vol. 462: P. Gérardin, Construction de Séries Discrètes p-adiques. »Sur les séries discrètes non ramifiées des groupes réductifs déployés p-adiques«. III, 180 pages. 1975.

Vol. 463: H.-H. Kuo, Gaussian Measures in Banach Spaces. VI, 224 pages. 1975.

Vol. 464: C. Rockland, Hypoellipticity and Eigenvalue Asymptotics. III, 171 pages. 1975.

Vol. 465: Séminaire de Probabilités IX. Proceedings 1973/74. Edité par P. A. Meyer. IV, 589 pages. 1975.

Vol. 466: Non-Commutative Harmonic Analysis. Proceedings 1974. Edited by J. Carmona, J. Dixmier and M. Vergne. VI, 231 pages. 1975.

Vol. 467: M. R. Essén, The Cos $\pi\lambda$ Theorem. With a paper by Christer Borell. VII, 112 pages. 1975.

Vol. 468: Dynamical Systems – Warwick 1974. Proceedings 1973/74. Edited by A. Manning. X, 405 pages. 1975.

Vol. 469: E. Binz, Continuous Convergence on C(X). IX, 140 pages. 1975.

Vol. 470: R. Bowen, Equilibrium States and the Ergodic Theory of Anosov Diffeomorphisms. III, 108 pages. 1975.

Vol. 471: R. S. Hamilton, Harmonic Maps of Manifolds with Boundary. III, 168 pages. 1975.

Vol. 472: Probability-Winter School. Proceedings 1975. Edited by Z. Ciesielski, K. Urbanik, and W. A. Woyczyński. VI, 283 pages. 1975.

Vol. 473: D. Burghelea, R. Lashof, and M. Rothenberg, Groups of Automorphisms of Manifolds. (with an appendix by E. Pedersen) VII, 156 pages. 1975.

Vol. 474: Séminaire Pierre Lelong (Analyse) Année 1973/74. Edité par P. Lelong. VI, 182 pages. 1975.

Vol. 475: Répartition Modulo 1. Actes du Colloque de Marseille-Luminy, 4 au 7 Juin 1974. Edité par G. Rauzy. V, 258 pages. 1975. 1975.

Vol. 476: Modular Functions of One Variable IV. Proceedings 1972. Edited by B. J. Birch and W. Kuyk. V, 151 pages. 1975.

Vol. 477: Optimization and Optimal Control. Proceedings 1974. Edited by R. Bulirsch, W. Oettli, and J. Stoer. VII, 294 pages. 1975.

Vol. 478: G. Schober, Univalent Functions – Selected Topics. V, 200 pages. 1975.

Vol. 479: S. D. Fisher and J. W. Jerome, Minimum Norm Extremals in Function Spaces. With Applications to Classical and Modern Analysis. VIII, 209 pages. 1975.

Vol. 480: X. M. Fernique, J. P. Conze et J. Gani, Ecole d'Eté de Probabilités de Saint-Flour IV–1974. Edité par P.-L. Hennequin. XI, 293 pages. 1975.

Vol. 481: M. de Guzmán, Differentiation of Integrals in R^n. XII, 226 pages. 1975.

Vol. 482: Fonctions de Plusieurs Variables Complexes II. Séminaire François Norguet 1974–1975. IX, 367 pages. 1975.

Vol. 483: R. D. M. Accola, Riemann Surfaces, Theta Functions, and Abelian Automorphisms Groups. III, 105 pages. 1975.

Vol. 484: Differential Topology and Geometry. Proceedings 1974. Edited by G. P. Joubert, R. P. Moussu, and R. H. Roussarie. IX, 287 pages. 1975.

Vol. 485: J. Diestel, Geometry of Banach Spaces – Selected Topics. XI, 282 pages. 1975.

Vol. 486: S. Stratila and D. Voiculescu, Representations of AF-Algebras and of the Group U (∞). IX, 169 pages. 1975.

Vol. 487: H. M. Reimann und T. Rychener, Funktionen beschränkter mittlerer Oszillation. VI, 141 Seiten. 1975.

Vol. 488: Representations of Algebras, Ottawa 1974. Proceedings 1974. Edited by V. Dlab and P. Gabriel. XII, 378 pages. 1975.

Vol. 489: J. Bair and R. Fourneau, Etude Géométrique des Espaces Vectoriels. Une Introduction. VII, 185 pages. 1975.

Vol. 490: The Geometry of Metric and Linear Spaces. Proceedings 1974. Edited by L. M. Kelly. X, 244 pages. 1975.

Vol. 491: K. A. Broughan, Invariants for Real-Generated Uniform Topological and Algebraic Categories. X, 197 pages. 1975.

Vol. 492: Infinitary Logic: In Memoriam Carol Karp. Edited by D. W. Kueker. VI, 206 pages. 1975.

Vol. 493: F. W. Kamber and P. Tondeur, Foliated Bundles and Characteristic Classes. XIII, 208 pages. 1975.

Vol. 494: A Cornea and G. Licea. Order and Potential Resolvent Families of Kernels. IV, 154 pages. 1975.

Vol. 495: A. Kerber, Representations of Permutation Groups II. V, 175 pages. 1975.

Vol. 496: L. H. Hodgkin and V. P. Snaith, Topics in K-Theory. Two Independent Contributions. III, 294 pages. 1975.

Vol. 497: Analyse Harmonique sur les Groupes de Lie. Proceedings 1973–75. Edité par P. Eymard et al. VI, 710 pages. 1975.

Vol. 498: Model Theory and Algebra. A Memorial Tribute to Abraham Robinson. Edited by D. H. Saracino and V. B. Weispfenning. X, 463 pages. 1975.

Vol. 499: Logic Conference, Kiel 1974. Proceedings. Edited by G. H. Müller, A. Oberschelp, and K. Potthoff. V, 651 pages 1975.

Vol. 500: Proof Theory Symposion, Kiel 1974. Proceedings. Edited by J. Diller and G. H. Müller. VIII, 383 pages. 1975.

Vol. 501: Spline Functions, Karlsruhe 1975. Proceedings. Edited by K. Böhmer, G. Meinardus, and W. Schempp. VI, 421 pages. 1976.

Vol. 502: János Galambos, Representations of Real Numbers by Infinite Series. VI, 146 pages. 1976.

Vol. 503: Applications of Methods of Functional Analysis to Problems in Mechanics. Proceedings 1975. Edited by P. Germain and B. Nayroles. XIX, 531 pages. 1976.

Vol. 504: S. Lang and H. F. Trotter, Frobenius Distributions in GL_2-Extensions. III, 274 pages. 1976.

Vol. 505: Advances in Complex Function Theory. Proceedings 1973/74. Edited by W. E. Kirwan and L. Zalcman. VIII, 203 pages. 1976.

Vol. 506: Numerical Analysis, Dundee 1975. Proceedings. Edited by G. A. Watson. X, 201 pages. 1976.

Vol. 507: M. C. Reed, Abstract Non-Linear Wave Equations. VI, 128 pages. 1976.

Vol. 508: E. Seneta, Regularly Varying Functions. V, 112 pages. 1976.

Vol. 509: D. E. Blair, Contact Manifolds in Riemannian Geometry. VI, 146 pages. 1976.

Vol. 510: V. Poènaru, Singularités C^∞ en Présence de Symétrie. V, 174 pages. 1976.

Vol. 511: Séminaire de Probabilités X. Proceedings 1974/75. Edité par P. A. Meyer. VI, 593 pages. 1976.

Vol. 512: Spaces of Analytic Functions, Kristiansand, Norway 1975 Proceedings. Edited by O. B. Bekken, B. K. Øksendal, and A. Stray VIII, 204 pages. 1976.

Vol. 513: R. B. Warfield, Jr. Nilpotent Groups. VIII, 115 pages. 1976

Vol. 514: Séminaire Bourbaki vol. 1974/75. Exposés 453 – 470. IV 276 pages. 1976.

Vol. 515: Bäcklund Transformations. Nashville, Tennessee 1974 Proceedings. Edited by R. M. Miura. VIII, 295 pages. 1976.